WJEC
CBAC

MODULAR BIOLOGY

Units BI4 & BI5

Gareth Rowlands

A LEVEL

WJEC A Level Modular Biology
Units BI 4 & BI 5

Published by the Welsh Joint Education Committee
245 Western Avenue, Cardiff CF5 2YX

First published 2001

Printed by Hackman Printers Ltd
Clydach Vale, Tonypandy, Rhondda, CF40 2XX

ISBN: 1 86085 487 7

Introduction

Students are often uncertain as to what is required of them in preparation for the exams

This booklet:

- ✓ provides guidance for students specifically studying WJEC A2 Biology.
- ✓ is a study guide and **you should build on the information provided by using a text book** and **your lesson notes**.
- ✓ gives some indication of the main points to be noted.
- ✓ gives some guide as to the depth of treatment required.

An active role in your studies

- In order to revise effectively it is important to take an **active role** in the process. Passive reading through your notes has limited value. The guide suggests '**Action points'** where you:
- ✓ draw labelled diagrams.
- ✓ draw graphs.
- ✓ analyse result tables.
- ✓ highlight key terms and provide definitions.
- ✓ construct comparison tables.
- ✓ describe processes using flowcharts/flow diagrams.
- This does not mean that these activities are the only ones that you should carry out in preparation for an exam. They are merely suggestions. There may, for example, be a number of other drawings that you should consider.

Revision hints

Examiners are aware that candidates are often not able to effectively show the examiner what they know.

To overcome this problem it is necessary to:

- ✓ devise a structured **revision programme**. Revise regularly - repetition is an important tool in learning and in preparing for exams. Revise each topic as soon as it is completed. Success is achieved through consistent work rather than attempting to commit everything to memory at the last minute.
- ✓ **understand basic principles** e.g. by learning definitions.
- ✓ **revise all topics**. Each unit will contain a number of structured questions with one essay question. It is not a good idea to work out which topics are 'going to come up'. Examiners take great care that the questions cover every topic.
- ✓ produce **annotated diagrams** – these are a good way of revising!
- ✓ familiarise yourself with **past exam questions**.
- ✓ **express yourself clearly and concisely** – frequently students, who know a reasonable amount of biology, don't do themselves justice in an exam because of their poor exam technique. Much of this is to do with the problem of language. It can make the difference between a good grade and an average grade. You need to be able to read, interpret, memorise, understand and convey information concisely in order to answer exam questions successfully.
- ✓ understand the **key words** that are used in exam questions.

It is hoped that the guide may prove of real value to the student.

The author is a chief examiner for the WJEC and has written a further revision guide for the new specification 2001 together with a question book.

Revision express “A-level study guide Biology” (Pearson Education) (ISBN 0-582-43177-8)
Revision express “AS Fast-track Biology” (Pearson Education) (ISBN 0-582-43233-2)

Both publications complement the information in this booklet and provide an extension to the activities and information outlined here. This booklet, which emphasises the WJEC specific material, is largely, though not exclusively, based on the study guide.

(Discounted Revision Express publications are available direct from Pearson Education, Edinburgh Gate, Harlow, Essex CM20 2JE)

Gareth Rowlands, July 2001

Websites

http://ncbe.reading.ac.uk.

http://vector.cshl.org

http://www.newscientist.com/nsplus/insight/gmworld/gmfood.html

http://jic.bbsrc.ac.uk/exhibitions/bio-future/index.htm

http://www.greenpeace.org.uk

http://www.wildlifetrust.org.uk

http://helios.bto.ed.ac.uk/bto/microbes/microbes.htm

http://www.mrc.ac.uk

http://www.wellcome.ac.uk

www.revision-express.com

www.wjec.co.uk

4.1 The role of ATP in metabolism

The importance of chemical energy in biological processes

- Energy cannot be created or destroyed.
- Chemical energy is contained within food substances.
- Energy may be converted from one form to another.
- Green plants are able to convert light into chemical energy. All living organisms are able to convert chemical energy to other energy forms.

The structure of ATP

ATP is a nucleotide consisting of an organic base, adenine, a five carbon sugar, ribose, and a sequence of three phosphate groups linked together.

❑ ***Action*** Draw the general structure of ATP.

The importance of ATP

- **Energy is released** when ATP is broken down to ADP and phosphate and this is an **exergonic** reaction. The hydrolysis of ATP to ADP is catalysed by the enzyme **ATP synthetase** and the removal of the terminal phosphate yields 30.7 kJ mol^{-1} of free energy.
- **Energy is required** to combine ADP and phosphate to form ATP and this is an **endergonic** reaction.

The addition of phosphate to ADP (**phosphorylation**) requires 30.7 kJ mol^{-1} of energy. ATP is therefore a means of transferring free energy from energy rich compounds to cellular reactions requiring it. There are two forms of phosphorylation:

✓ **oxidative phosphorylation**, which occurs during cellular respiration in all aerobic cells.
✓ **photosynthetic photophosphorylation,** which occurs during photosynthesis in chlorophyll-containing cells.

The uses and advantages of ATP as a source of energy

There are several advantages in having ATP as an intermediate source of energy compared with glucose. These include:

- Only one enzyme is needed to release energy from ATP while many are needed in the case of glucose.
- ATP releases energy in small amounts when and where needed whereas glucose contains large amounts of energy that may not be needed immediately.
- ATP provides a common source of energy for many different chemical reactions, increasing efficiency and control by the cell. ATP is in fact known as the "universal energy currency in living organisms".

ATP is not stored, it has to be synthesised as required. The rate of synthesis keeps pace with demand. A metabolically active cell may require 2 million ATP molecules per second!
ATP provides the necessary energy for:

✓ active transport
✓ muscle contraction
✓ synthesis of materials within cells
✓ nerve transmission.

❑ ***Action*** Distinguish between the terms 'oxidative phosphorylation' and 'photosynthetic photophosphorylation'.

Summary

The oxidation of carbohydrate, protein and fat during catabolism liberates large amounts of energy. However, direct oxidation of these molecules with oxygen would release the energy as heat that could not be used by cells. The main method of energy re-distribution involves the breakdown of large molecules using many chemical steps. Some of these steps involve inorganic phosphate and the phosphorylation of ADP to ATP. This traps a lot of the energy in a form that can be released in a controlled manner to drive energy-requiring processes such as growth, synthesis of molecules, active transport and mechanical work. ATP therefore serves as a common medium of exchange between catabolic and anabolic processes.

❑ ***Action*** This summary provides an example of an effective way of revising. Read a section of your notes, close your file and write out a summary.

The chemiosmotic theory of ATP synthesis

Consider the following points:

- Hydrogen atoms split up into protons (H^+) and electrons.
- Many reactions in respiration involve the release of electrons at a high energy level.
- These electrons are collected by electron carrier molecules which pass them to electron carriers at lower energy levels.
- The energy released is used to form ATP from ADP.
- Reactions in which electrons are transferred from one reactant to another are called oxidation-reduction reactions, often abbreviated to 'redox' reactions.
- When a substance is oxidised it loses electrons and when it is reduced it gains electrons.
- Anabolism involves the reduction of molecules and hence the intake of energy.
- Catabolism involves the oxidation of molecules and a consequent release of energy.
- Molecules with a lot of hydrogen and little oxygen have the greatest potential to provide energy.
- Dehydrogenases are enzymes that catalyse the removal of hydrogen from their substrate and transfer it to a carrier.

The chemiosmotic theory proposes that the energy for ATP synthesis comes from the **electrochemical gradient** across a membrane. Mitochondria have an inner membrane that is folded to form **cristae** that are lined with stalked bodies.

Within the inner membrane there is a mechanism which actively transports H^+ ions (protons) from the matrix into the space between the inner and outer membranes of the mitochondrion. This creates an electrochemical gradient of **protons** across the inner membrane.

The process occurs as follows:

- The oxidation of glucose in the matrix is coupled to the reduction of a carrier.
- The reduced carrier passes an **electron** to an associated pump in the membrane. (The 'pumps' in this context are large enzyme molecules that traverse the membrane.)
- The carrier thus becomes reoxidised and releases protons into the matrix. The charged pump then pumps protons into the space between the inner and outer membranes.
- Electrons with a slightly reduced energy level are transferred by a carrier to the next pump. (a series of carriers and pumps is known as the **electron transport system**).
- A high concentration of protons builds up in the inter-membrane space because the inner membrane is impermeable to protons, except at certain sites.
- One of these sites is the enzyme ATP synthetase.
- The protons are able to flow back into the matrix via channels formed by ATP synthetase molecules in the stalked bodies.

- This flow acts as a driving force to combine ADP and inorganic phosphate to synthesise ATP.
- The electrons from the last pump are taken up by oxygen, which combines with protons to form water. This is catalysed by the enzyme oxidase.

The same principle applies for ATP synthesis in chloroplasts and mitochondria.
In chloroplasts the build up of protons is inside the membrane whereas in mitochondria the build up of protons is outside the membrane.

❑ ***Action***

1. Draw a labelled diagram of a mitochondrion.
2. Draw a diagram to explain the synthesis of ATP as described above.
3. Rephrase the list of points about the chemiosmotic theory on page 4 as essay style paragraphs.
4. Explain what is meant by the terms 'redox reaction', 'electrochemical gradient', 'carrier' and 'pump'.

4.2 Aerobic respiration

Respiration is a series of enzyme catalysed reactions which release energy from organic molecules in order to synthesise ATP. Aerobic respiration, which involves the participation of oxygen, consists of four distinct but linked stages, glycolysis, the link reaction, Krebs Cycle, and the electron transport chain.

Glycolysis

Glycolysis takes place in the cytoplasm and oxygen is not required.

- The **glucose** molecule is first **phosphorylated** to make it more reactive by the addition of two molecules of ATP to form **hexose phosphate**.
- The 6-carbon hexose phosphate is split into two molecules of **triose phosphate** (3-carbon sugar).
- Each 3-carbon sugar is converted to **pyruvic acid**.
- Two of the steps transfer sufficient energy for the synthesis of ATP yielding a total of four ATP molecules. Since these are formed directly from phosphate compounds in the process it is known as **substrate level phosphorylation.**
- Each 3-carbon sugar initially donates two hydrogen atoms to reduce the carrier molecule NAD. If oxygen is available NADH/H^+ enters the mitochondrion and brings about the synthesis of 6 ATP molecules.

❑ ***Action*** 1. Draw a flowchart that summarises glycolysis.
2. Work out the net gain of ATP molecules in glycolysis.

Link reaction

- Pyruvic acid diffuses from the cytoplasm to the mitochondrial matrix.
- Pyruvic acid (3-carbon) is converted to 2-carbon acetate with a loss of hydrogen to NAD, and carbon dioxide.
- The acetate then combines with coenzyme A to form **acetyl coenzyme A**.

Krebs Cycle

1. The function of the Krebs cycle is a means of liberating energy from carbon bonds to provide ATP and reduced NAD (and FAD), with the release of carbon dioxide.
2. Reduced NAD (and FAD) deliver the hydrogen to the electron transport system in the inner mitochondrial membrane so acting as triggers for this system.

- The acetyl CoA then enters the Krebs cycle by combining with a 4-carbon acid, to form a 6-carbon compound, citric acid; the CoA is regenerated.
- The 6-carbon compound undergoes reactions during which carbon dioxide and hydrogen atoms are removed and the acetate fragment from the glucose molecule is completely broken down. The 4-carbon compound is regenerated via 6-carbon and 5-carbon intermediates, and is ready to combine with another molecule of acetyl CoA.
- Two of the steps involve **decarboxylation,** and four of the steps involve **dehydrogenation**.
- The hydrogen atoms produced are collected by two different carriers with the formation of three molecules of NADH/H^+ (reduced NAD) and one molecule of $FADH_2$ (reduced FAD).
- For each turn of the cycle, one ATP molecule is produced by substrate level phosphorylation and fourteen are produced via the electron transport chain (**oxidative phosphorylation**).

Electron transport system

- Although the glucose molecule has been completely oxidised by the end of the Krebs cycle, much of the energy is in the form of hydrogen atoms which are attached to the hydrogen acceptor molecules NAD and FAD.
- The hydrogen atoms split into their protons and electrons. The protons enter the space between the inner and outer membrane of the mitochondrion by the **proton pump** mechanism (see the previous section) while the electrons pass along the carriers and associated pumps, known as the **electron transport chain**, located within the inner membrane.
- It is the flow of protons back into the matrix via the stalked granules (due to their high concentration in the intermembrane space) that acts as the driving force to synthesise ATP.
- These protons then recombine in the matrix with the electrons and the hydrogen atoms so formed then combine with oxygen to form water.
- If NADH is the initial acceptor, for each pair of hydrogen atoms involved enough energy is released for the synthesis of three molecules of ATP. If FADH replaces NADH as the first carrier only two molecules of ATP are produced.

Note that:

- *The names of the proton pumps and electron carriers in the electron transport system are **not** required.*
- *The names of the intermediates of glycolysis and Krebs cycle are **not** required.*
- *You are **not** required to know the names of all the enzymes involved at each stage but you should know that dehydrogenase enzymes catalyse the removal of hydrogen atoms and decarboxylases catalyse the removal of carbon dioxide.*

❑ ***Action***

1. Draw a diagram that summarises aerobic respiration. (Include only the names of those chemicals mentioned in bold on page 6 and use only terms such as 6C, 5C and 4C in the Krebs cycle.)
2. Insert the points of removal of carbon dioxide and hydrogen atoms.
3. Show on your diagram that for each pair of hydrogen atoms evolved, three molecules of ATP are produced (except for one of the carrier systems in the Krebs cycle where FAD is the first acceptor).
4. Total the *net* gain of ATP molecules from one molecule of glucose. (Don't forget that 2ATPs are used in the phosphorylation of the glucose molecule and that you must double the number of ATPs produced by the Krebs cycle because for each glucose molecule broken down, the cycle turns twice – once for each molecule of pyruvic acid produced.) *This is dealt with in detail in the next section.*
5. Make a list of key terms and write definitions for each.

Anaerobic respiration, energy budgets

Anaerobic respiration occurs in some microorganisms and in some tissues, e.g. muscle cells which may be temporarily deprived of oxygen. However, without oxygen, oxidative phosphorylation will not take place and so the yield of ATP molecules will be greatly reduced. You need to be able to compare and explain the differences in yields of ATP from aerobic and anaerobic respiration.

Anaerobic respiration

In the absence of oxygen:

- Only the first stage of respiration, glycolysis, can take place.
- The reduced NAD (and FAD) cannot be reoxidised and therefore made available to pick up more hydrogen, and so the link reaction and the Krebs cycle cannot take place.
- The yield of ATP is greatly reduced.

You have learnt from the previous section that in the matrix protons recombine with electrons to form hydrogen atoms which then combine with oxygen to form water. Since most of the yield of useful energy by metabolism comes from the transfer of hydrogen atoms or electrons, oxygen must be available to accept the hydrogen atom. If oxygen is absent the hydrogen atoms are not removed at the end of the respiratory chain and will accumulate. While oxygen only performs this function at the end of the many stages of respiration, it is nevertheless vital as it drives the whole process.

- In vertebrate muscle cells **lactic acid** is produced. The pyruvate itself becomes the hydrogen acceptor.
- In higher plants and yeast, **alcohol** is the product. This is also known as alcoholic fermentation. Ethanol (acetaldehyde), formed by the decarboxylation of pyruvate, is the hydrogen acceptor.
  A considerable amount of energy remains trapped in ethanol and lactic acid.

Energy budget

Aerobic respiration

Remember that:
Usually 3 ATP molecules are produced every time 2 hydrogen atoms pass through the carrier system.
The energy yield for each molecule of reduced NAD is three ATP
The energy yield for each molecule of reduced FAD is two ATP
As there are two pyruvate molecules formed for each glucose molecule all figures for the link reaction and Krebs cycle must be doubled (x 2).

Glycolysis
2 ATP used (in the phosphorylation of glucose).
4 ATP produced directly (by substrate level phosphorylation).
6 ATP produced (from 2 pairs of H atoms picked up by NAD which may enter the mitochondrion and pass along the electron transport system).
8 ATP = net gain

Link reaction
(2 x) 3 ATP
6 ATP

Krebs Cycle (2 x) 1 ATP formed directly (by substrate level phosphorylation).
(2 x) 11 ATP from electron transport chain
(three pairs of hydrogen atoms via NAD yields 9 ATP; one pair via FAD yields 2 ATP)
24 ATP = net gain from Krebs Cycle

38 ATP molecules = Total net gain

The efficiency of aerobic respiration.

Each ATP molecule will yield 30.7 kJ of energy
Therefore total energy available is 38 x 30.7 = 1166.6 kJ
The complete oxidation of glucose yields 2880 kJ
Therefore efficiency of aerobic respiration $= \frac{38 \times 30.7 \times 100}{2880} = 40.3\%$
Not all the energy of the glucose molecule is captured in ATP as there is a loss of energy as heat energy.

Anaerobic respiration

Energy yield = 2 ATP (glycolysis only)
Therefore total energy available = 61.4 kJ
Efficiency of anaerobic respiration $= \frac{61.4 \times 100}{2880} = 2.13\%$
This process is just over 2% efficient.

Alternative respiratory substrates

In addition to glucose, both fats and proteins may, under certain circumstances, be used as respiratory substrates, e.g. long distance running and starvation.

Fat respiration

- Fat is a good energy store and is used as a respiratory substrate when carbohydrate is in short supply. It has to be split into its constituent molecules of **glycerol and fatty acids**, first by hydrolysis and then the glycerol is converted into a 3-carbon sugar which enters the Krebs Cycle via **triose phosphate** yielding ATP molecules.
- The long fatty acid chains molecules are split into 2-carbon fragments which enter the pathway as acetyl CoA. Very large numbers of ATP molecules are built up during the process, the precise number depending on the length of the hydrocarbon chain of the fatty acid, e.g. the complete oxidation of stearic acid yields nearly four times as many ATP molecules as given by the oxidation of a single glucose molecule.

Protein

In prolonged starvation the tissue protein is mobilised to supply energy and whenever dietary energy supplies are inadequate the protein component of the food is diverted for energy purposes.
Protein is very rarely used as a respiratory substrate, usually only when all reserves of carbohydrate and fat have been used up. The protein is hydrolysed into its constituent **amino acids** and then it is deaminated in the liver. The amino group is converted into urea and

excreted and the residue is converted to either acetyl CoA, pyruvic acid or some other Krebs Cycle intermediate, and oxidised.

Notice that most of the metabolic pathways lead to acetyl CoA which is a kind of crossroads in metabolism. As well as its formation during the oxidation of carbohydrates it is formed during the oxidation of fats and proteins and represents a common pathway by which the products of all three are fed into the Krebs cycle.

- ***Action***
  1. Define the terms aerobic and anaerobic.
  2. Draw a simplified scheme showing how carbohydrate, fat and protein metabolism are interconnected.
  3. Describe similarities in mitochondrial and chloroplast membranes in terms of providing a proton gradient: proton pumps, ATP synthetase, electrochemical gradient, high energy electrons fueling the pump.

- ***Class practical*** Carry out an experiment to show how dehydrogenase activity may be demonstrated in respiring tissue using artificial hydrogen acceptors which change colour as they are reduced by accepting hydrogen.

4.3 Photosynthesis

Photosynthesis takes place in the chloroplasts which are found in the mesophyll cells and guard cells of green leaves. In the chloroplasts the energy of sunlight is trapped by the pigment chlorophyll. Since photosynthesis is a biochemical process involving many enzymes a suitable temperature is necessary, in addition to light, carbon dioxide and water.

Photosynthetic pigments

The role of the photosynthetic pigments is to absorb light energy and to convert it to chemical energy. There are two types of pigments in flowering plants:

- The **chlorophylls** – these absorb mainly in the red and blue-violet regions of the spectrum. There are a number of different chlorophylls with chlorophyll a and b being the most common.
- The **carotenoids** – these absorb blue-violet light. They are thought to act as **accessory pigments**. There are two main types of carotenoids, the carotenes and the xanthophylls.

Chromatography is a means of separating one type of molecule from another.
You would be expected to identify the pigment spots on the chromatogram by their colours and positions by measuring Rf values.

$$Rf = \frac{\text{distance moved by spot}}{\text{distance moved by solvent}}$$

❑ ***Class practical*** Separate the pigments from a leaf using the technique of paper chromatography.
(Hint: you may find that a leaf square (2mm) pressed onto the filter paper gives better results than the' loading' method.)

Absorption and action spectra

Chlorophyll absorbs only certain wavelengths, reflecting others.
An **absorption spectrum** indicates that chlorophyll absorbs wavelengths in the blue and red parts of the spectrum. However, this does not indicate whether the wavelengths are actually used. This can be shown by plotting an **action spectrum** of the amount of carbohydrate synthesised by plants exposed to different wavelengths of light. Since the absorption spectrum and the action spectrum for chlorophyll show a close correlation it suggests that this pigment is responsible for absorbing the light used in photosynthesis.

❑ ***Action*** Draw the action spectrum for photosynthesis and the absorption spectrum for chlorophyll on the same graph.

Biochemistry of photosynthesis

Photosynthesis is a three-stage process. You will first need to consider how light energy is captured by the plant. The light dependent reactions result in the formation of ATP and reduced NADP which are then used in the light independent stage in the reduction of carbon dioxide and the synthesis of organic compounds.

Light harvesting

Light energy is captured by the plant using a mixture of pigments, including chlorophyll. Within the **thylakoid** membranes of the chloroplast, chlorophyll molecules are arranged along with their accessory pigments into groups of several hundred molecules. Each group is called an **antenna complex**. Special proteins associated with these pigments help to funnel photons of light energy entering the chloroplast. Chlorophyll b is an accessory pigment molecule which passes on its energy to chlorophyll a, which is a primary pigment molecule and is known as the **reaction centre**. When light strikes this molecule an electron is raised to a higher energy level, thus setting up a flow of electrons.

There are two types of photosynthetic units, **Photosystem I (PSI)** which is at a higher energy level than **Photosystem II (PSII)**.

❑ ***Action***

1. Revise the structure of the chloroplast by drawing a labelled diagram.
2. Draw one thylakoid in detail showing the position of the reaction centres of photosystems I and II within the thylakoid membrane, together with an associated stalked particle.

Light dependent stage

The light dependent stage occurs in the thylakoids of the chloroplasts and involves the splitting of water by light (**photolysis**). In the process ADP is converted to ATP.

This is known as **photophosphorylation**.

Textbooks differ in their use of the term 'reduced NADP'. Some use $NADPH_2$, others $NADPH^+H^+$ (the correct terminology used by the Biochemical Society in their booklet 'Biochemistry across the school curriculum'). Throughout this section the term 'reduced NADP' will be used.

During the light dependent stage electron transport occurs as follows.

- The absorption of light energy, by certain chlorophyll molecules in the reaction centre, causes the displacement of 'excited' (high energy) electrons which are accepted by other molecules called **electron acceptors**. This is an **oxidation-reduction** process in which the pigment molecule, the electron donor, is oxidised and the acceptor molecule is reduced.
- As electron transport occurs hydrogen ions are pumped from the stroma into the thylakoid discs which make up each granum. These H^+ flow out and down a gradient of both concentration and electrical potential difference with the production of ATP.
- Electrons from chlorophyll are passed into the light independent reaction via reduced NADP. These electrons are replaced by electrons from the water molecule. Since these electrons are not recycled back into the chlorophyll this method of ATP production is called **non-cyclic photophosphorylation**.

The process occurs as follows.

✓ Light energy absorbed by PSII causes the displacement of an excited electron to a higher energy level. The electrons are received by an electron acceptor and passed via a carrier through a proton pump, from which the electrons are carried to PSI. The energy lost by the electrons is captured by converting ADP to ATP.

✓ Light energy absorbed by PSI boosts the electrons to an even higher energy level and are picked up by another electron acceptor. The electrons which have been removed from the chlorophyll are replaced by the electrons from a water molecule which dissociates into protons and oxygen. The protons from the water molecule combine with the electrons from the second electron acceptor and these reduce NADP.

ATP can also be generated by **cyclic photophosphorylation**.
Some electrons from PSI return to the chlorophyll directly via the electron carrier system, forming ATP in the process. These electrons are recycled, harnessing energy from light and generating ATP. No reduced NADP is produced.

Consider these points:

- You do **not** need to know the names of the electron acceptors.
- Non-cyclic photophosphorylation can be represented by the Z scheme (so called because the zig-zag route of the electrons in the diagram resembles a Z on its side). This summary diagram will help you to understand the process but try also to consider the position of the pigment systems within the thylakoid membrane.
- The loss of electrons is a form of oxidation.
- The hydroxide ion from the splitting of water, is a source of electrons for photosystem II.
- Molecules of water split into hydrogen and oxygen because electrons are removed to replace the electrons lost in the chlorophyll in the photosystems. Light is responsible only indirectly for splitting water.
- The reduction of the co-enzyme NADP is brought about by the addition of electrons and hydrogen ions. This takes place in the stroma therefore maintaining a proton gradient across the thylakoid membrane.
- Hydrolysis of water occurs in the thylakoid space; the reduction of NADP occurs in the stroma; the former produces protons and the latter removes them. Therefore they both contribute to maintaining the electrochemical gradient across the membrane as described in section 1.
- The light dependent stage produces reduced NADP, ATP and oxygen.

❑ ***Action*** Draw an annotated diagram to illustrate the light dependent stage.

Light independent stage

The light independent stage occurs in the stroma of the chloroplast and involves many reactions each catalysed by a different enzyme. The reactions use the products of the light dependent stage, ATP as a source of energy, and reduced NADP as the source of the reducing power to reduce carbon dioxide and synthesise hexose sugar.

The sequence of events in this stage of photosynthesis was worked out by Calvin and his associates using C^{14}, a radioisotope of carbon, and the unicellular algae, *Chlorella*. It is known as the Calvin Cycle.

- A five-carbon acceptor molecule, **ribulose bisphosphate (RuBP)**, combines with carbon dioxide (catalysed by the enzyme Rubisco) forming an unstable six-carbon compound.
- The six-carbon compound immediately splits into two molecules of a three-carbon compound called **glycerate-3-phosphate** (GP).
- GP is **phosphorylated** by ATP and then **reduced** by reduced NADP (from the light dependent stage) to **triose phosphate**.

- Some of this three-carbon sugar can be built up into glucose phosphate and then into starch by condensation.
- In order that the cycle continues, the majority of the triose phosphate formed enters a series of reactions, driven by ATP which results in the regeneration of RuBP.

❑ ***Action*** Draw a diagram to summarise the light independent stage.

Products of photosynthesis

The plant must synthesis all organic materials from the intermediates and products of photosynthesis.

From triose phosphate glucose, lipids and amino acids (with the addition of nitrogen obtained from nitrates) may be manufactured. (You are **not** required to provide chemical details.)

The end products of carbohydrate digestion are all monosacharides. The final stage of carbohydrate digestion is intracellular, as disaccharides are absorbed by the plasma membrane of the epithelial cells before being broken down into monosaccharides.

❑ ***Action*** 1. For the process of digestion complete a table with the following headings: Organ of secretion; site of action; contents; effects.
2. Draw a detailed portion of part of the wall of the small intestine showing Brunner's glands opening into crypts of Lieberkühn.

Absorption

Absorption follows digestion and takes place mainly in the small intestine. The surface available for absorption is greatly increased by the presence of villi and microvilli. Because energy is required for active absorption, the epithelial cells also contain large numbers of mitochondria.

❑ ***Action*** Draw a detailed section through a villus and show the regions of absorption of the different products of digestion.

- **Glucose and amino acids** are absorbed across the epithelium of the villi by a combination of diffusion and active transport. They pass into the capillary network that supplies each villus. The blood from the venules, which contains the dissolved food, eventually reaches the hepatic portal vein and and is carried to the liver.
- **Fatty acids and glycerol** are passed into the lacteal. This is a blindly ending lymph capillary found in the centre of each villus. Fatty acids and glycerol are transported in the lymphatic system which ultimately opens into the blood stream at the thoracic duct.

Water and mineral salts are absorbed from the colon along with vitamins secreted by microorganisms in the colon. By the time it reaches the rectum, indigestible food is in a semi-solid condition. It consists of residues of undigested cellulose, bacteria and sloughed cells and passes along the colon to be egested as faeces. This process is called defaecation.

Uses of the products of digestion

With the completion of the processes of digestion and absorption the soluble food products are carried in the bloodstream to the tissues for **assimilation** or to provide **energy**.

- Glucose is absorbed from the blood by cells, for energy release in respiration, and any excess is stored in the liver as glycogen and in fat cells.
- Amino acids are absorbed for protein synthesis, excess cannot be stored so are deaminated whereby the removed amino groups are converted to urea and the remainder to carbohydrate and stored.
- Lipids are used for membranes and hormones, excess are stored as fat.

4.5 Microbiology

Consider reviewing the relevant parts of Unit BI 1 before beginning this section. Remember that microorganisms include fungi and viruses as well as bacteria. However this section of the specification requires a reference to bacteria unless otherwise stated.

Bacteria

Bacteria are the smallest cellular organisms and are prokaryotes. The kingdom includes both heterotrophic and autotrophic forms

There are several **ways of classifying bacteria**.

- One way is by their **shape**. Bacteria may be:
- ✓ **Bacillus** or rod shaped.
- ✓ **Cocci** or spherical.
- ✓ **Spirillum** or corkscrew shaped.

The shape of bacteria is due to their rigid cell wall which has a unique structure. It contains peptidoglycan (murein) consisting of molecules of polysaccharide cross-linked by amino acid side chains. The cross-linking provides strength and the wall protects against osmotic lysis.

❑ ***Action*** Review the structure of a typical bacterial cell.

Further differentiation is often possible according to the way bacteria tend to be grouped e.g. in pairs; forming chains; in clusters. The filamentous bacteria are of particular interest because most are photosynthetic. This group, formerly called blue-green algae, are now known as Cyanobacteria.

- Another method of classification is based on the variation in **nature of bacterial cell walls.** The gram stain technique distinguishes two types of bacteria:
- **Gram negative bacteria** have chemically complex walls where the glyco-protein is supplemented by large molecules of lipopolysaccharide which protect the cell. They do not retain dyes like crystal violet, e.g. *Salmonella* sp.
- **Gram positive bacteria** lack the lipopolysaccharide in their walls and do retain crystal violet. These are more susceptible to antibiotics and the enzyme, lysozyme, which occurs in human tears, than gram negative bacteria, e.g. *Staphylococci*, *Streptococci.*

Bacteria reproduce asexually by **binary fission** although certain bacteria can also reproduce sexually.

Although some bacteria are pathogens and some cause deterioration of stored food, it is important to remember that many are beneficial to humans.

Conditions necessary for growth

Microorganisms quickly reproduce given a suitable environment. In the laboratory they can be grown on a wide variety of substrates providing they are supplied with suitable physical conditions, nutrients and water. Organisms vary in their requirements and usually grow over a range of temperatures and pH values, with an optimum within the range.

Microorganisms require the following conditions for growth:

- **Nutrients** – in the laboratory nutrients are supplied in nutrient media and include carbon, usually organic such as glucose; nitrogen, organic and inorganic form; growth factors such as vitamins and mineral salts.
- **Temperature** – as all growth is normally regulated by enzymes the range of 25-45°C is favourable for the majority of bacteria. The optimum for mammalian pathogens is around 37° C.
- **pH** – most bacteria are favoured by slightly alkaline conditions (pH 7.4) *(whereas fungi prefer neutral to slightly acid conditions).*
- **Oxygen** – many microorganisms require oxygen for metabolism and are termed **obligate aerobes**. Some, while growing better in the presence of oxygen, can nevertheless survive in its absence; these are called **facultative anaerobes**. Others cannot grow in the presence of oxygen and are called **obligate anaerobes**.

Growth of microorganisms by means of fermenters

Biotechnology may be defined as the industrial application of biological processes. For many of these processes the key steps are conducted in a fermenter vessel. You should understand the difference between batch and continuous fermentation and how industrial processes involve the need for aseptic entry of material and the control of conditions to ensure optimum yield of product.

Fermenter design

- A pure culture of an organism is needed for the formation and harvesting of a pure product during and after growth in a fermenter vessel. The organism must be supplied with suitable conditions for growth and without competition for maximum efficiency.
- The vessel should be sterilised beforehand and an appropriate sterile medium used. During use the vessel openings must be protected from contamination by filters and aseptic conditions and handling are required to maintain purity.
- Forced aeration may be needed, for maximum growth of aerobes, which may also mix the culture to improve contact with nutrients. Mixing may be improved by a separate mixer.
- Temperature monitoring and control are required to maintain constant conditions and water-jackets remove excess heat produced during the culture process.
- Commercially, sophisticated monitors are used to improve control of temperature and pH, and air inlets may use spargers or other devices to improve aeration.
- The products of fermentation may be:

✓ a dissolved substance that accumulates in the medium e.g. the antibiotic penicillin is produced by the fungus *Penicillium*. In this case the used culture medium, after filtering, is retained and processed.

✓ the microorganism itself e.g. single-cell protein (mycoprotein) manufacture. In this case the medium may be recycled and the solid processed.

❑ ***Action*** Draw an annotated diagram of a batch culture fermenter.
Explain the biological importance of the various components.

Batch versus continuous cultivation

- **Batch fermentation** – as the name suggests, for a particular reaction to occur, growth is allowed to continue up to a specific point at which the fermenter is emptied and the product extracted. The fermenter has to be cleaned and sterilised in readiness for the next batch.
- **Continuous fermentation** – once the fermenter is set up the used medium and products are continuously removed. The raw materials are added throughout.

Antibiotic production

- For the commercial production of penicillin, the fungus *Penicillium notatum* is grown in a batch culture and the antibiotic is produced after the growth phase, when glucose is depleted. The mycelium is removed by filtration of the culture fluid and the antibiotic purified from the residual liquid.
- The primary metabolism is the norm, when the fungus is respiring glucose to release energy and producing an increase in its own biomass. Antibiotic production is a **secondary metabolism**; i.e. the antibiotic is produced at a period in the life of the fungus when there is a change away from its optimum conditions. This reflects the need for the organism, when free living, to reduce competition when food sources are depleted.

Growth and culture techniques

The growth of a culture of individual cells follows the same pattern as the growth of any population i.e. lag phase, exponential phase, stationary phase and death phase.

Aseptic technique

Bacteria (*and fungi*) are cultured (grown) on, or in, media that are designed to supply the cell with all their nutritional requirements. **Aseptic** techniques (also known as sterile techniques) in which the apparatus and equipment are kept free of microorganisms, are used to prevent contamination of bacterial cultures and the surrounding environment.

- **To prevent** the **contamination** of pure cultures and apparatus by bacteria **from** the **environment**:
- ✓ Sterilise all apparatus and media before use to prevent initial contamination;
- ✓ Handle cultures carefully and use equipment such as sterile loops to prevent subsequent contamination.
- **To prevent contamination to the environment** by the bacteria being used in experiments:
- ✓ Sterilise the work surface before and after an experiment using a disinfectant, e.g. Lysol used as a 3% solution.
- ✓ Use the **correct handling techniques** to prevent the contamination of personnel and the immediate environment by the organisms being cultured;
  e.g. when carrying out the process of inoculation:
  - Grasp the culture bottle in one hand; remove the cap with the little finger of the other hand – do not place the cap down on the work surface.
  - Flame the mouth of the bottle for 2 or 3 seconds.
  - Pass the inoculating loop through a flame until red hot.
  - Lift the lid of the petri dish just enough to allow entry of the inoculating loop.
  - Secure the petri dish lid with adhesive tape. Use two pieces of tape to fasten the lid, but do not seal all the way round (as this could create anaerobic conditions and encourage the growth of possible pathogenic microorganisms).
  - Incubate at around 25°C (cultures should not be cultured at 37°C as this is an ideal temperature for the growth of many pathogenic species).
  - Do not open petri dishes after incubation.
- ✓ In a laboratory the preferred method of sterilisation is to use an **autoclave**. This is a sealed container in which glass and metal equipment is heated at 121°C in steam under pressure for 15 minutes after the required pressure has been reached. This ensures that any resistant endospores are destroyed e.g. gram positive bacteria such as *Clostridium* are resistant to boiling.
  Disposable materials, such as plastic Petri dishes, can be sealed inside autoclavable plastic bags and placed in a dustbin after having been autoclaved.
- ✓ Radiation e.g. gamma rays, is used commercially to sterilise plastic equipment.

❑ ***Action***
1. Make a list of all the terms used in microbiology e.g. inoculation, media, sterilisation etc. and write out the definitions of each term.
2. Describe how you would safely obtain a pure culture of one species of bacterium from a culture containing a mixed bacterial suspension in a nutrient medium.

❑ ***Class practical*** Gram staining and microscopic investigation of yoghurt.

Methods of measuring growth

- ➢ The size of a population of microorganisms in liquid culture may be measured by counting cells directly, or by taking some indirect method such as the turbidity (cloudiness) of the culture.

  Direct cell counts may be divided into:
- ✓ **total counts**, which include both living and dead cells.
- ✓ **viable counts**, which count living cells only.

- ➢ In practice, it is never possible to count whole populations of microorganisms. Instead, the cells in a very small **sample** of culture are counted, and the result multiplied up to give a population density in organisms per cm^3 of culture. Even then, the population density is likely to be so high that cell counts are usually made in known dilutions of the culture, usually in 10-fold steps. This is known as **serial dilution**.

There are several different methods of measuring growth:

- Rough estimates of growth rates can be made by regularly measuring the diameter of a bacterial or fungal colony as it spreads from a central point to cover the surface of a solid growth medium.
- Changes in cell density can be measured by the following methods:
- ✓ Using a **haemocytometer** which measures **total counts** (both living and dead cells).
- ✓ By plating known volumes onto a solid nutrient medium, viable counts can be made by counting the number of colonies which grow. Here it is assumed that one cell divides to form a colony. This makes no allowance for clumping of cells so may cause an underestimate of numbers.
- ✓ Photometry – a **colorimeter** can be used to measure light intensity and turbidity as cell numbers increase. (You may find this method in textbooks but it is not on the syllabus and will not be tested.)

❑ ***Class practical***

1. Carry out an investigation to investigate the number of bacteria in fresh and stale milk using the techniques of serial dilution, plating and counting colonies. (This will give a viable count.)
2. If available, use a haemocytometer to measure the total count of a sample. Alternatively, use a computer simulation. (Remember that cells are counted as being 'in' a square if they do not touch the sides, or if they touch the top or right-hand side.)

4.6 Pathogens and human disease

Another aspect of microbiology is the effect of the different pathogenic organisms and their control. Bacteria, viruses and fungi can be pathogens of humans, their crops and their domestic animals and the consequences can be of great economic significance.

Disease transmission

- Many of the important diseases in humans are caused by **pathogenic** bacteria or viruses. These may be transmitted from one person to another through the air, via contaminated food and water, by direct contact or with the aid of vectors.
- Vectors are organisms that carry pathogens from one organism to another e.g. mosquitoes act as vectors for the Protoctistan *Plasmodium*, which causes malaria.
- A disease that can be passed from host to host is termed an **infectious** disease.
- Diseases spread by direct contact are described as **contagious**.
- A **carrier** is infected by a disease organism, showing no symptoms, but can pass the disease on to another individual.
- The place where a pathogen is normally found is its **reservoir**. This may be in humans or another animal and may be a source of infection.
- A disease which is always present at low levels in an area is described as **endemic.**
- An **epidemic** is where there is a significant increase in the usual number of cases of a disease often associated with a rapid spread.
- **Air-borne transmission** is also called droplet infection, and is the main way in which diseases of the lungs and respiratory passages are spread. The microorganisms are expelled in tiny droplets of saliva and mucus as a result of sneezing, coughing, or even just breathing. Bacterial diseases spread this way include whooping cough and tuberculosis, while viral infections include influenza and measles.
- Transmission via food and water – these can become contaminated with bacteria if sanitation and hygiene are poor. Salmonella food poisoning is spread in or from the undercooked meat of infected animals while diseases such as cholera and typhoid are mainly spread through the contamination of water by the untreated faeces of infected individuals.

Other terms you need to know

- A **toxin** is a chemical produced by a microorganism which causes damage to its host.
- An **antigen** is any substance that when introduced into the blood or tissues induces the formation of antibodies, or reacts with them if they are already present.
- An **antibody** is a substance produced by lymphocytes (white blood cells), in the presence of a specific antigen, that can combine with that antigen to neutralise, inhibit or destroy it.
- A **vaccine** uses non pathogenic forms or products of microorganisms to stimulate an immune response which confers protection against subsequent infection.
- **Antibiotics** are substances produced by microorganisms which affect the growth of other microorganisms.
- Antibiotic **resistance** is where a microorganism which should be affected by an antibiotic is no longer susceptible to it.
- **Antigenic types** indicate organisms with the same or very similar antigens on the surface. Such types are sub groups or strains of a microbial species which may be used to trace infections. They are usually identified by using antibodies from serum.

Selected diseases, their control methods and treatment

*For the diseases **Salmonellosis, Cholera, Influenza** and **Malaria** you should know:*

✓ *The name of the organism that causes the disease.*
✓ *The tissue affected and symptoms (in brief).*
✓ *The source of infection.*
✓ *The mode of transmission.*
✓ *The prevention and control methods (including vaccines).*

Bacterial diseases

- **Salmonella** food poisoning is caused by a gram negative rod-shaped bacterium.

✓ Poisonous substances, called **toxins**, produced by the bacteria, affect the gut lining causing diarrhoea and vomiting.
✓ It is commonly found in animal intestines and may be transferred to meat at animal slaughter. The bacterium multiplies during storage of infected food.
✓ Good hygiene practices are essential to avoid food poisoning. These include thorough cooking, storage in cool conditions, separation of cooked and uncooked meat, the prevention of contamination from carriers. Analysis of antigenic types may enable the tracing of the source of an infection.
✓ Antibiotic treatment is possible but not usual to prevent the build up of resistance. A vaccine is not available as there are over 2000 antigenic types.

- **Cholera** is caused by a gram-negative bacterium, endemic in some areas of the world.

✓ Toxins affect the gut lining causing watery diarrhoea leading to severe dehydration and possibly death. Humans can act as a reservoir and as carriers.
✓ This disease is often spread by infected water in which the organism does not multiply but may become easily widespread.
✓ Protective measures are a clean water supply, the safe disposal of sewage and the hygienic preparation of food.
✓ Powerful antibiotic therapy is possible but treatment is largely by rehydration. Vaccine (killed organism or possibly genetically engineered) may provide temporary protection.

Antibiotic action

✓ Antibiotics either prevent growth, bacteriostatic, or kill bacteria, bactericidal, depending on which aspect of bacterial metabolism is affected.
✓ Those used medically affect bacterial metabolism but do not interfere with the host cell metabolism.
✓ Penicillin affects the formation of cross linkages in the cell wall, the wall is weakened so when osmotic changes occur, the cells lyse. Consequently penicillin is more effective against Gram positive organisms than Gram negative due to the difference in the structure of the cell wall.
✓ Some antibiotics affect metabolic processes common to most bacteria, such as protein synthesis, so are effective against a broader range of bacteria.

Virus diseases

Several highly significant human diseases are caused by viruses including rubella, shingles, measles, poliomyelitis, influenza and HIV.

Influenza is caused by a tiny virus consisting of a central strand of RNA, coated in protein, surrounded by an outer lipoprotein coat.

- ✓ It infects cells of the lining of the upper respiratory tract, causing sore throat, cough, feverishness.
- ✓ Transmission is by droplet infection.
- ✓ Prevention includes quarantine and hygiene but it is a difficult mode of spread to control.
- ✓ Antibiotic treatment has no effect therefore symptoms are treated as necessary. (Viruses are not affected due to the absence of metabolic pathways.)
- ✓ Vaccines are available but due to the number of antigenic types (strains) and the regular emergence of new ones, they are not always effective.

Protoctistan disease

Malaria is a major human disease caused by the parasite, *Plasmodium,* of which there are two species within which there are many antigenic types. The life cycle occurs in the human body and the female mosquito acts as a secondary host and vector. It is endemic in some sub-tropical regions.

- ✓ The infection is acquired by a human when an adult female mosquito, already infected with the malarial parasite, inserts its mouthparts into the individual's skin. The parasite infects the liver where it multiplies and then the red blood cells, which eventually burst to release parasites causing severe bouts of fever. Infected human or animal blood is taken in by a feeding mosquito which transmits the parasite to a new host.
- ✓ Prevention relies on a knowledge of the vector and the life cycle of both the vector and parasite in order to exploit their weak points. Malaria has been combatted in three ways:

1. **Preventing the mosquito from biting humans** by:
- ✓ using nets.
- ✓ wearing protective clothing.
- ✓ using insect repellants.

2. **Eliminating the vector** (the mosquito) by:
- ✓ draining swamps (breeding grounds of the mosquito).
- ✓ spraying the breeding grounds with diesel oil – the oil is taken into the breathing tubes of the larvae and kills them.
- ✓ using insecticides, e.g. DDT. (Resistance is an increasing problem.)
- ✓ using biological control methods - using fish (guppies) to eat larvae, bacteria to infect larvae, sterilisation of males to reduce breeding.

3. **Attacking the parasite**:
- ✓ Using drugs, e.g. quinine – this method merely reduces the chances of infection. *Plasmodium* is affected by drugs when outside the cells in the blood but these have limited effectiveness and have side effects; antibodies also are only effective against the parasite when outside body cells so limiting the target stages for a vaccine.
- ✓ Vaccines have proved difficult to develop because the malarial parasite mutates and there are different antigenic types.

❑ ***Action***
1. Summarise the above information about the four diseases by constructing a table with the headings- *organism, source, transmission and prevention/control.*
2. Make a list of the advantages and disadvantages of the methods of controlling malaria.

4.7 Protection against disease

Defence mechanisms

The human body operates a number of defence mechanisms against disease:

- **Natural barriers** such as the skin, skin flora and blood clotting to seal wounds along with protection by ciliated mucous membranes, lysozyme and stomach acid.
- Resistance to disease also depends on the **general health and diet**, for instance, a deficiency of vitamin C leads to weakened connective tissue causing open wounds.
- The **skin flora** (harmless microorganisms that live on the skin surface), which is never completely removed, prevents colonisation by pathogens which are relatively easy to remove by washing.
- Localised defence involves **phagocytosis** and **inflammation** which localises any break in the barrier and destroys invading organisms.
- The **immune system** involves the recognition of foreign material and the production of antibodies which help to destroy it. It is also possible to induce an individual to produce antibodies without them suffering from the disease. This is the basis of immunisation or vaccination.

The immune system

- **Immunity** is the ability of an organism to resist disease. It is a specific systemic response acquired during a lifetime.
- An **antigen** is any substance that when introduced into the blood or tissues induces the formation of antibodies, or reacts with them if they are already present. Antigens are normally large, complex molecules such as proteins and other polymers. Antigens are called self-antigens if they occur on the surface of cells in the body and non-self if they are from 'foreign' organisms.
- **Antibodies** are proteins (immunoglobulins) produced in the body to combine with foreign antigens. An antibody is a Y-shaped molecule made up from a large pair and a small pair of polypeptide chains. Each antibody is specific for a particular antigen. An antibody is a substance produced by lymphocytes (white blood cells) in the presence of a specific antigen, that can combine with that antigen to neutralise, inhibit or destroy it.

Immunity

Immunity involves two initial processes:

- The recognition of foreign material when it enters the body and the mobilisation of cells and antibodies capable of removing the foreign material, or antigens, quickly and effectively.
- The formation of **memory cells**, which carry out the **secondary immune response**.

There are two systems of immunity in mammals: a **cell-mediated immune response** and **a humoral immune response**. Two types of lymphocytes are involved, both of which have receptor sites on their membranes for recognising antigens. Both types are derived from **stem cells** in the bone marrow. The lymphocytes that migrate to the **thymus gland** for maturation are called **T-lymphocytes** or T-cell, whereas those which remain in the bone marrow to complete their development are called **B-lymphocytes** or B-cells.

Humoral immune response
This response involves B-lymphocytes (B-cells).
There are many thousands of different B-lymphocytes, each has different immunoglobulins on its cell surface. When an antigen enters the blood it combines with the few B cells carrying the appropriate immunoglobulin. These divide rapidly by mitosis forming a clone of plasma cells in the lymph node. Most of these cells produce antibodies but some produce memory cells. The plasma cells live only a few days but can synthesise and secrete vast quantities of specific antibody molecules.
The memory cells persist for a long time, sometimes for life. They are responsible for the secondary immune response and confer active immunity against the specific antigen. If that antigen is encountered again, the memory cells can recognise it and stimulate the immediate production of massive quantities of antibody.

Cell-mediated immune response
This response involves T-lymphocytes (T-cells).
The thymus gland is active from birth until a mammal is weaned. During this time it causes lymphocytes to mature and become 'immunologically competent', and capable of synthesising new receptor molecules and incorporating them into the plasma membrane. T-lymphocytes leave the thymus and circulate in the blood and body fluids. If a T-lymphocyte meets an antigen for which it has a receptor site, it is stimulated to divide many times by mitosis, forming a clone. Each cell in the clone can attach to a complementary antigen and destroy it.
Among these dividing cells three functional sub-populations occur (in addition to the formation of memory cells):

- ✓ T killer cells cause lysis of the target cells, e.g. they destroy virus infected or cancer cells.
- ✓ Helper T cells which activate B lymphocytes to initiate an antibody response;
- ✓ T suppressor cells suppress other cells in the immune system, e.g. by turning off antibody production when the antigen is no longer present.

Types of immunity

There are two types of immunity:

- **Naturally acquired immunity** – this may occur passively or actively.
- ✓ Natural passive immunity may be due to the transfer of antibodies from mother to fetus across the placenta or from mother to new-born offspring via colostrum, the first secretion of the mammary gland. The immunity is only temporary since no memory cells have been formed.
- ✓ Natural active immunity is achieved as a result of exposure to infection. The body manufactures its own antibodies in response to the presence of antigens on the infectious agent and also forms specific memory cells. If the same agent is encountered again, it can be flooded with antibodies and eliminated before it causes disease.

- **Artificially acquired immunity** – also occurs passively or actively.
- ✓ Artificial passive immunity results from the injection of ready-made antibodies into the body, and again, since there are no memory cells, it is only temporary, e.g. tetanus antitoxin. It is useful as a preventative measure for diseases that are difficult to immunise against, such as tetanus and diphtheria. It may also be used as a treatment for certain diseases, such as rabies, where infection has already occurred and which is too dangerous to leave to the body's natural immune system.
  Protection is short lived because it is recognised as non-self and destroyed.

- ✓ Artificial active immunity - is achieved by injecting vaccine into a healthy individual, e.g. against Rubella. The body is stimulated to produce antibodies and memory cells against the antigen in the vaccine, and thus acquires immunity to subsequent infection by that disease organism.

❑ ***Action*** Distinguish between the following terms:
antigen and antibiotic
cell-mediated and humoral immune response
B cells and T cells
active and passive immunity
primary and secondary immune response.

How do vaccines work?

Put simply:
The vaccine contains antigens, but is not capable of causing disease. When it is injected into the body, lymphocytes recognise the antigens as 'non-self', and make antibodies to destroy them. This is the **primary response**. Memory cells then remain in the blood for a long time: these cells are capable of recognising the same antigens if it ever enters the body again, and producing the same antibodies immediately.
If at some time in the future, the pathogen enters the body, memory cells recognise the antigens on its surface and antibodies are produced immediately. These destroy the pathogen before it makes the person ill: this is the **secondary response**.

In terms of antibody production:
Exposure to an antigen is followed by a latent period, then the primary response, when the antibody concentration increases before decreasing again.
If the antigen is reintroduced or if it persists for the duration of the primary response, it stimulates a secondary response.
This is characterised by requiring a lower level of antigen, shorter latent phase, higher concentrations of antibody are produced and longer duration of peak antibody concentration.
Each subsequent exposure acts as a booster until maximum levels are reached.
This pattern of response is mimicked by immunisation programmes such as for Rubella.

There are three main differences between the primary and secondary response:

- ✓ There is a delay before antibodies are produced in the primary response, but in the secondary response they are produced immediately.
- ✓ Larger amounts of antibodies are produced in the secondary response.
- ✓ The antibodies remain in the blood for longer during the secondary response.

❑ ***Action*** Draw a graph showing the effect of immunisation on antibody production.

The relative effectiveness of vaccination programmes

You should consider the relative effectiveness of specific vaccination programmes as well as the ethical considerations involved.

- Vaccination has made a major contribution to reducing the incidence of certain diseases in the population. Its effectiveness is improved the greater the proportion of the population vaccinated. So the incidence is reduced. Eventually, as there is no animal reservoir, the disease is eradicated locally and then possibly globally, as in the case of smallpox. International co-operation is essential for this to occur. The original decision to vaccinate only teenage girls against german measles has been questioned as boys would act as a reservoir of infection, so should also be vaccinated.

- Smallpox vaccination has been effective because the organism does not have a high mutation rate so the vaccine remained effective, unlike influenza which mutates frequently and causes periodic epidemics due to lack of resistance in the population and lack of effective long term vaccination.

- Reports of possible side effects in a few cases of vaccination against "childhood" diseases reduces the uptake of vaccination leading to the increase of incidence of the disease, so increasing the risk of contact. Often a disease will have serious side effects in a small number of cases.

- Public information and education are therefore important to ensure rational decision making along with the identification of any contra-indications against vaccination in individual cases to minimise personal risk.

- Increased travel increases the risk of infections spreading globally. Influenza is spread by droplet infection, so is difficult to control; has a high mutation rate, so vaccines are not effective; is a viral infection, so there is no effective curative treatment; may have animal and bird reservoirs to act as a source of infection.

4.8 Applications and contemporary issues

Recent advances in biotechnology have brought about the discovery of a variety of new ways of treating disease. Monoclonal antibodies are used to diagnose disease, to locate and isolate other molecules and in diagnostic testing. Fermenters have enabled the large-scale production of genetically engineered products such as insulin, myco-protein and antibiotics. You need to consider the mode of action of the latter, their effects on microorganisms which cause disease and the problems associated with their overuse

The commercial production of penicillin

The production of penicillin by batch culture in a fermenter has been dealt with on page 21. A stirred-tank fermenter is inoculated with a culture of *Penicillium notatum* or *Penicillium chrysogenum* and the fungus is grown under optimum conditions: 24°C, good oxygen supply, slightly alkaline pH. Penicillin production typically reaches a maximum at around four days. Production ceases at six days, at which point the contents of the fermenter are drained off and the liquid filtrate containing penicillin is extracted. After sterilization the fermenter is ready to receive the next batch.

❑ ***Action*** Draw an annotated diagram showing the stages in the commercial production of penicillin.

Monoclonal antibodies

- Traditional methods of producing antibodies involving laboratory animals have resulted in impure preparations. Natural antibodies are a mixture because several antibodies may be formed in response to an invading bacterium. A monoclonal antibody is one that responds to only one foreign antigen. It is the product of cells called hybridomas, made from mice myeloma cells.
- In monoclonal antibody production, spleen cells are exposed to a single, known antigen. They produce a single, specific antibody to it. These spleen cells are fused with myeloma cells and the result is a clone producing a specific antibody. The hybridoma cells are grown by tissue culture and the antibody they secrete is collected and concentrated.

❑ ***Action*** Illustrate the production of monoclonal antibodies by means of a flowchart.

Monoclonal antibodies have a number of **applications**:

✓ Antibodies may be tagged with a marker the detection of which is used for the diagnosis of specific conditions e.g. **pregnancy-testing kits**. Home pregnancy-testing kits make use of immobilised antibodies on a urine dipstick to detect traces of human chorionic gonadotrophin (hCG), a hormone released from the placenta. Antibodies 'tagged' with blue latex combine with the hormone to produce a readily visible result.

✓ **Drug targetting** - to act as 'magic bullets'.
This involves linking anti-cancer drugs with monoclonal antibodies which are attracted to cancer cells. This enables these drugs to be delivered to specific target cells, such as tumours, so improving effectiveness and reducing side effects.

✓ To treat a range of infections.

✓ In the diagnosis of disease, e.g. to identify people who are carriers for some genetic diseases.

✓ To separate a particular chemical from a complex mixture.

Antibiotics

Antibiotics are substances which in low concentration inhibit the growth of microorganisms. They can be divided into two groups:

- Antibiotics effective against a wide range of pathogenic bacteria are called 'broad spectrum' antibiotics e.g. chloramphenicol, tetracyclines. These antibiotics affect widespread microbial processes such as protein synthesis.
- Antibiotics effective against a limited range of pathogens are called 'narrow spectrum' antibiotics e.g. penicillin. These antibiotics affect more specific processes such as cell wall formation.

Antibiotic resistance

Resistance to a chemical poison is the ability of an organism to survive exposure to a dose of that poison which would normally be lethal to it.

- The use of antibiotics to treat disease in the human population has had obvious benefits. However, in agriculture, although the prophylactic use of small quantities of antibiotics in animal feed has led to healthier, faster growing animals, this widespread increased use of antibiotics has led to the development of resistance among many species of bacteria.
- Resistance arises by mutations occurring randomly within populations of organisms which then confers an advantage in the presence of that antibiotic. This may be the ability to produce an enzyme which breaks down the antibiotic. e.g. Some bacteria have developed an enzyme, penicillinase, which renders penicillin ineffective. In the presence of penicillin non-resistant forms are destroyed. There is a selection pressure favouring the resistant types. The greater the quantity and frequency of penicillin use, the greater the selection pressure.
- The problem has been made worse by the discovery that resistance can be transmitted between species. There is evidence that the resistance may be passed from one organism to another on plasmids, during conjugation (sexual reproduction). This means that a disease-causing organism can become resistant to a given antibiotic even before the antibiotic is used against them. MRSA (methycillin resistant *Staphylococcus aureus*) has developed a resistance to several antibiotics. This organism originated in Australia and within ten years has spread world-wide.A sample of *Pseudomonas aeruginosa* has been identified as being resistant to all clinical antibiotics.
- Such organisms are becoming a more common problem and the antibiotics available to treat them very limited in number.

Vaccination programmes

This topic has been covered on pages 30 and 31.

5.1 Variability and genetics

Mendel carried out the first documented scientific investigation of inheritance and formulated two laws, which are basic to the science of genetics. You need to understand the principles of monohybrid and dihybrid Mendelian inheritance.

Genes and alleles

In 1866 Gregor Mendel suggested that the characteristics of organisms were determined by 'units' which were handed on from generation to generation. Later these units were identified as **genes** which were carried on and transmitted by chromosomes. A gene is the basic unit of inheritance.

Genes consist of DNA and have three main characteristics:

- They can separate and combine.
- They can mutate.
- They code for the production of specific polypeptides.

The definition of a gene based on function is 'one gene being the portion of a chromosome which codes for one polypeptide'.

Alleles are alternative forms of genes occupying a similar gene-position (or 'locus') on homologous chromosomes. If a gene determines a particular inherited characteristic, the alleles which make up the gene may exist in two forms. For example, in a gene which determines fur colour in mice, the two alternative alleles may be for black and white fur. 'Black' and 'white' are the two alleles for the 'fur colour' gene.

For any one locus on a chromosome, there are theoretically three different allele combinations.

- ✓ **heterozygous** – having different alleles for a given gene. i.e. a dominant allele and a recessive allele are present together. Each of these alleles is carried on a different chromosome, within a pair of homologous ('identical') chromosomes.
- ✓ **homozygous** dominant – having the same two dominant alleles present for a given gene.
- ✓ **homozygous** recessive - having the same two recessive alleles present for a given gene.

In the simplest situations, a particular characteristic is controlled by a single gene. If an organism is heterozygous for this gene, the dominant allele will determine the form in which the characteristic is actually expressed. The outcome (i.e. the **phenotype**) of a heterozygous condition (i.e. the **genotype**) will therefore be the same as that for a homozygous dominant condition.

Genetic terms

- **Dominant** allele - of the pair of alleles in a monohybrid cross, the one that always produces an effect on the appearance of the organism when present is the dominant allele. It is usually represented by a capital letter e.g. T (for tall).
- **Recessive** allele – the allele that produces an effect only when present as an identical pair is the recessive allele e.g. t.
- **Genotype** – the combination of alleles found in an individual.
- **Phenotype** – the appearance of an organism, determined by the genotype.
- **Homozygous** – if both alleles are the same e.g. TT, tt.
- **Heterozygous** – if the pair of alleles are dissimilar e.g. Tt.
- **F1** – first filial generation; **F2** – second filial generation.

❑ ***Action***

1. Distinguish between the following terms: gene and allele; dominant and recessive; genotype and phenotype; homozygous and heterozygous.
2. Draw a labelled diagram of a homologous pair of chromosomes showing alternative alleles for a single gene, e.g. fur colour in mice.

Mendel's first Law

Mendel's early experiments were based on selecting pea plants of two varieties which showed clearly separable characteristics such as tall and dwarf plants and round and wrinkled seeds. From his results he formulated the law of segregation, which states that:
'the characteristics of an organism are determined by factors which occur in pairs. Only one member of a pair of factors can be represented in a single gamete'.
The inheritance of a single pair of contrasting characters is known as **monohybrid inheritance**.
In monohybrid crosses, two heterozygous individuals will produce offspring with a a phenotypic ratio of three dominant to one recessive.

Instructions for carrying out a genetic cross

(Once you have practised a number of crosses it is very easy to miss out stages or explanations. This may make your explanations impossible for others to follow. In an exam, even if you achieve the expected outcome, you may not gain full credit. So always carry out these instructions in their entirety).

1. Look carefully at the information you are given in the question, e.g. about phenotypes of the parents, whether they are homozygous or heterozygous.
2. Choose suitable symbols for the alleles, unless they are provided in the question.
    a) Choose a single letter to represent each characteristic.
    b) Choose the first letter of one of the contrasting features.
    c) If possible, choose a letter in which the higher and lower case forms differ in shape as well as size.
    d) Let the higher case letter represent the dominant feature and the lower case letter the recessive one.
3. Represent the parents with the appropriate pairs of letters. Label them clearly as 'parents' and state their phenotypes.
4. State the gametes produced by each parent. Circle the gametes and label them clearly.
5. Use a matrix, called a Punnett square, to show the results of the random crossing of the gametes.
6. State the phenotype of each different genotype and indicate the numbers of each type.
7. Make sure that you have actually answered the question.

❑ ***Action*** Draw a genetic diagram to show the outcome of a cross between tall pea plants crossed with short pea plants. Use suitable symbols, to include F1 and F2 generations.

Test cross

The 'backcross' or test cross is a method used in genetics to determine whether a particular dominant characteristic observed in an organism is determined by one or two dominant alleles. For example, a mouse with black fur could be homozygous dominant (BB) or heterozygous (Bb). The appearance of the black fur (phenotype) is identical in both cases.
The backcross consists of a 'cross' (i.e. cross fertilization) between an unknown 'dominant' genotype and a known 'recessive' genotype. (The double recessive phenotype has a known genotype because only one allele combination can produce it.)

❑ ***Action*** Draw two genetic diagrams to show how a backcross can reveal the genotype of a plant having a dominant phenotype (tall).

Mendel's second Law

Mendel also investigated what happened if he used plants which differed by having two pairs of contrasting characters. In one of his experiments he investigated the inheritance of seed shape (round v. wrinkled) and seed colour (green v. yellow). He crossed homozygous plants with round and yellow seeds with homozygous plants with wrinkled and green seeds. When plants grown from these seeds were self-pollinated the seeds produced were of four different types of shape and colour of seed coat. The four types were:

- ✓ round yellow
- ✓ round green
- ✓ wrinkled yellow
- ✓ wrinkled green

He also found that these types were in the ratio of approximately 9:3:3:1, proportions which are now known as the dihybrid ratio. This led him to formulate his second law, stating that:
'either one of a pair of contrasted characters may combine with either of another pair'.
With our present knowledge of genetics this statement can be rewritten as:
'each member of an allelic pair may combine randomly with either of another pair'.

❑ ***Action*** Draw a genetic diagram to show the outcome of a cross between round green seeds and wrinkled yellow seeds. Use suitable symbols, to include F1 and F2 generations.

Probability

The presentation of a breeding experiment, such as that shown in the 'Action' above, is a prediction of the likely outcome. What is actually observed may not agree precisely with the prediction. Consider the situation where a coin is tossed 100 times. It would be expected to land heads on 50 occasions and tails on 50 occasions. In practice it would be unusual to obtain this result the first 100 attempts. If the coin lands 60 heads and 40 tails is this due to a chance deviation from the expected result or is the coin biased in some way?
The following description is to aid your understanding that the results of conventional genetic crosses are due to chance. You would not be required to reproduce this description in a theory exam.

If two unbiased coins are tossed there are four possible combinations:

HH or HT or TH or TT.

- Rule of addition:
  Consider one of the coins. When this is tossed it is certain to give either H or T. Thus the probability of getting either one result or another result is obtained by *adding* their independent probabilities. The probability of getting H is ½ and the probability of getting T is also ½.

  Thus the probability of getting one or the other is

  ½ + ½ = 1 or 100% certainty.

- Rule of multiplication:
  Now consider both coins. The probability that there will be H on both the first and the second coin is obtained by multiplying the two independent probabilities. In this case the probability of getting H on the first coin is ½ , and the probability of getting H on the second coin is also ½.
  Thus the probability of HH is

  ½ x ½ = ¼.

The **t-test** is a statistical test required *only* in the practical Unit BI3. The **null hypothesis** (negative hypothesis) assumes any differences in an investigation have arisen by **chance**. The role of the statistical test is to determine whether to accept or reject the null hypothesis. If it is rejected we can have confidence that any differences between two samples are statistically significant and have not occurred by chance.

In the cross between round green seeds and wrinkled yellow seeds described in the 'Action' above:

- the probability of the four alleles appearing in any of the F2 offspring is:
  round (dominant) ¾ yellow (dominant) ¾
  wrinkled (recessive) ¼ green (recessive) ¼

- The probability of combinations of alleles appearing in the F2 is as follows:

✓ round and yellow = ¾ x ¾ = 9/16
✓ round and green = ¾ x ¼ = 3/16
✓ wrinkled and yellow = ¼ x ¾ = 3/16
✓ wrinkled and green = ¼ x ¼ = 1/16

Not all the characteristics are controlled by single genes which behave independently, as was the case in Mendel's experiments. You need to know about incomplete dominance (monohybrid only) and also how sex is determined and how genes on sex chromosomes cause unexpected ratios.

Codominance

When alleles express themselves equally in the phenotype this is known as co-dominance. In most cases the heterozygote shows a phenotype intermediate between those of the two homozygotes.
Examples of co-dominance are:

- antirrhinums – red (RR), pink (Rr) and white (rr) flowers.
- shorthorn cattle – red (RR), roan (Rr) and white (rr) coat colour.

The genetic diagram for these crosses is the same as that illustrating Mendel's first law but in the F1 all individuals have the intermediate phenotype.

❑ ***Action*** Draw a genetic diagram to show the outcome of a cross between a homozygous red shorthorn bull and a white shorthorn cow. Use suitable symbols to include the F1.

(Where codominance is involved it is usual to use different letters to represent each allele, e.g. R to represent red and W to represent white. So the F1 would be shown as RW.)

Sex determination

Most sexually reproducing animals show two morphologically distinct types, male and female, which are associated with the chromosomes found in the two types. One pair, known as the **sex chromosomes**, are similar in one sex and dissimilar in the other. The non-sex chromosomes are known as **autosomes**. In humans the male has dissimilar chromosomes, called **X** and **Y**, whilst the female has two similar **X** chromosomes.

❑ ***Action*** Draw a simple genetic diagram to illustrate the determination of sex.

Sex-linked inheritance

Some alleles are carried on the X chromosome, so they are described as sex linked. (The Y chromosome is much smaller than the X and carries very few, if any, genes.) Therefore in the male any recessive genes carried on the X chromosome will express themselves in the phenotype. This is because they are unpaired and so there is no dominant gene present. This special form of inheritance is known as **sex-linkage**, an important feature of which is that the male cannot hand on the gene to his sons as they must receive the Y chromosome to become male. On the other hand, all his daughters must receive the recessive gene from him. Females who are heterozygous for sex-linked recessive traits are known as carriers and have a 50% chance of handing on the recessive to their sons. An example of a sex-linked trait in humans is **haemophilia**.

- Haemophilia is a potentially lethal condition. It is the result of an individual being unable to produce one of the many clotting factors. The inability of the blood to clot leads to slow and persistent bleeding.
- It is now possible to extract the particular clotting factor from donated blood allowing haemophiliacs to lead near-normal lives. (Although the risk of passing the disease on to their children remains.)
- This condition occurs almost exclusively in males.
- When the recessive allele occurs in males it expresses itself because the Y chromosome cannot carry any corresponding dominant allele.
- For the condition to arise in females requires the double recessive state and as the recessive allele is relatively rare in the population this is unlikely to occur.
- To obtain an affected female, the father must be affected and the mother either affected or a carrier.
- Heterozygous females are not themselves affected but are capable of passing the recessive allele to their offspring. They are therefore known as carriers.

If H is the allele for normal blood clotting
h is the allele for haemophilia
an individual could have the genotype

X^HX^H	female, normal
X^HX^h	female carrier
X^hX^h	female, haemophiliac
X^HY	male, normal
X^hY	male, haemophiliac

❑ ***Action***

1. Construct a genetic diagram to show the outcome of a cross between a man who does not have haemophilia and a woman who carries the allele for haemophilia. Describe the phenotypes of the offspring. What is the probability of a son being a haemophiliac? What is the probability of a daughter being a carrier?
2. Construct a genetic diagram to show the outcome of a cross between a carrier mother and a haemophiliac father.

Don't forget to include all the stages for carrying out a genetic cross listed on page 35.

Mutations

A mutation is a change in the genetic material and is one of the most important topics in genetics. You need to study point (gene) mutations and chromosome mutations and their effect as illustrated by sickle-cell anaemia and Down's Syndrome.

(You should link this topic with Unit 1.7, 'Meiosis and sources of variation'.)

A mutation is a change in the amount, arrangement or structure in the DNA of an organism.

- It may affect a single gene or a whole chromosome.
- Most mutations occur in somatic (body) cells.
- Only those mutations which occur in the formation of gametes can be inherited.
- Mutations are spontaneous random events which may provide a source of material for natural selection pressures and therefore evolution.
- Mutations rates are normally very small therefore mutation has less impact on evolution than other sources of variation.
- In general, organisms with short life cycles and more frequent meiosis, show a greater rate of mutation.
- The rate of mutations occurring can be increased by ionising radiation and mutagenic chemicals.

Mutagens and the mutation rate

Mutations happen naturally. However, scientists have found that the mutation rate is increased if organisms are exposed to mutagens. These are factors in the environment which increase the mutation rate. These include:

- X-rays, gamma radiation and UV light;
- chemicals, such as polycyclic hydrocarbons in cigarette smoke.

A mutagen which causes cancer is a carcinogen.

It is incorrect to say that mutagens cause mutations. Increased exposure to mutagens increases the *rate* of mutations occurring.

Mutations can happen in two ways:

- DNA is not copied properly before cell division.
  Sometimes mistakes are made in the copying process so that new chromosomes are faulty. Usually they are small mistakes, involving only one gene, so they are called **gene mutations** or point mutations. However, this can be a serious problem for the individual if a very important gene is affected.

- Chromosomes are damaged and break.
  If chromosomes break, they will normally repair themselves (the DNA will rejoin), but they may not repair themselves correctly. This can lead to large changes in the structure of the DNA, and may affect a large number of genes. These are called **chromosome mutations**.

Gene mutations

Any gene can mutate but rates vary from one gene to another within an organism. Gene mutations are changes in the base pairs within the genes. They can take the form of duplication, insertion, deletion, inversion or substitution of bases. Whatever the change, the result is the formation of a modified polypeptide.
A gene mutation (substitution) in the gene producing haemoglobin results in a defect called **sickle-cell anaemia**.
The replacement of just one base in the DNA molecule results in the wrong amino acid being incorporated into two of the polypeptide chains which make up the haemoglobin molecule. The abnormal haemoglobin causes red blood cells to become sickle-shaped, resulting in anaemia and possible death. Haemoglobin S is produced instead of normal haemoglobin by a single base chain that causes valine to be substituted for glutamic acid at the sixth position in the β globin chain. DNA codes for glutamic acid are CTT or CTC. Two of the codes for valine are CAT and CAC. In either case the substitution of A for T as the second base would bring about the formation of haemoglobin S.
The mutant gene is co-dominant. In the homozygous state the individual suffers the disease but in the heterozygous state the individual has 30-40% sickle cells, the rest are normal. This is called sickle-cell trait.

Chromosome mutations

These may affect either the number or the structure of chromosomes.

- ***Changes in numbers***

This is usually the result of errors occurring during meiosis.
Aneuploidy is the loss or gain of a **single** chromosome. It results in the formation of gametes with either one too many or one too few chromosomes. One pair of chromosomes has failed to separate at anaphase and so the pair has moved to one pole of the cell, this is **non-disjunction**. After fertilisation a zygote formed with an extra chromosome is said to be trisomic e.g. **Down's Syndrome** where chromosome number 21 is trisomic. The error occurs in the production of ova rather than sperm and the incidence of the mutation is related to the age of the mother.

(Polyploidy, involving changes in whole sets of chromosomes is not on the syllabus.)

- ***Changes in structure***

During chiasmata formation, mistakes arise when chromatids break at these points and rejoin with the corresponding pieces of chromatid on its homologous partner. (No details are required but this type of mutation is important as a source of variation.)

How can mutations cause a change in phenotype?

The genetic code, which ultimately determines an organism's characteristics, is made up of a specific sequence of nucleotides on the DNA molecule. Any change to one or more of these nucleotides, or any rearrangement of the sequence, will produce the incorrect sequence of amino acids in the protein it makes.
The protein made is often an enzyme, which may then be unable to catalyse a specific reaction. For example, a specific enzyme is necessary to convert a chemical precursor into the skin pigment, melanin. If a gene mutation results in the inability to produce this enzyme the organism will be unpigmented, i.e. an albino.

Why are mutations important?

Mutations are important because they increase variation in a population. Most mutations are harmful to the individual concerned. The effects of mutations are rarely beneficial.

- If a mutation is in a body cell, it may cause cancer. For example increased exposure to UV light is linked to skin cancer.
- If the mutation is in a gamete, it will not affect the individual producing the gamete, but will affect the zygote that develops from it, i.e. the offspring.
- These mutations cause sudden and distinct differences between individuals. They are therefore the basis of discontinuous variation.
- There are potential advantages from mutations that are beneficial and may increase variation. However, most mutations are recessive to the normal allele. A recessive mutant allele must await replication in the gene pool over many generations before chance brings recessive alleles together, resulting in their expression.

5.2 Variation and evolution

In the long term, if a species is to survive in a constantly changing environment and to colonise new environments, sources of variation are essential.

Types of variation

There are two types of variation:

- **Discontinuous variation** - normally controlled by a single gene. This gene may have two or more alleles. There are no intermediate types e.g. light and dark forms in some moth species.
- **Continuous variation**

A character within a population which shows a gradation from one extreme to another shows continuous variation e.g. height. It is influenced by the combined effect of a number of genes (polygenes). The effect of an individual gene may be small but their combined effect is large.

Origins of variation

Environmental effects

An organism will inherit genes, giving it a theoretical maximum size, but whether or not this is reached will depend upon nutrition during the growth period and other environmental factors. Thus, if organisms of identical genotype are subject to different environmental influences, they show considerable variety. Because these influences are varied, they are largely responsible for continuous variation in a population. This is known as **non-heritable variation.**
Variations due to the effect of the environment have little evolutionary significance as they are not passed from one generation to the next. Much more important to evolution are the inherited forms of variation which result from genetic changes.

Sexual reproduction

As a result of sexual reproduction, variation may be increased when the genotype of one parent is mixed with that of the other. The sexual process has three inbuilt methods of creating variety:

- The mixing of two different parental genotypes where cross-fertilization occurs.
- The random distribution of chromosomes during metaphase I of meiosis.
- The crossing over between homologous chromosomes during prophase I of meiosis.

(Review your notes from 1.7)

Mutations

Although these processes may establish a new combination of alleles in one generation it is mutations that generate long-lasting variation of a novel kind. However, as previously stated, the occurrence of a useful mutation is a very rare event.

Competition between organisms

Organisms must compete for limited resources of all kinds. For example, plants compete for light, space, mineral ions etc.; animals compete for food, shelter etc. *(Link with 2.5)*

- Competition between individuals of the same species (intra-specific competition) is the basis of the origin of species by natural selection.
- Competition between individuals of different species (inter-specific competition) is illustrated by predator-prey relationships.

Selection pressure

Selection, in the context of evolution, is the process by which organisms that are better adapted to their environment survive and breed, while those less well adapted fail to do so. These better adapted organisms are more likely to pass on their characteristics to succeeding generations. The organism's environment exerts a **selection pressure** and this determines the spread of any allele within the gene pool.

Gene pool

A population of organisms reproducing sexually contains a large amount of genetic variation called a **gene pool.** Each organism contains just one of the many possible sets of genes that can be formed from the pool. The gene pool remains stable if the environment is stable. However, if the environment changes some phenotypes will be advantageous and will be selected for, whilst others will be at a disadvantage and will be selected against. Thus a gene pool is constantly changing, some alleles becoming more frequent and others less frequent. In some circumstances alleles may be totally lost from the gene pool.
In large populations, assuming there is no selection, the proportion of dominant and recessive alleles of a particular gene remains constant. It is not altered by interbreeding.

(The Hardy-Weinberg principle is not specifically on the syllabus and so will not be required in an exam. However, your teacher may use it to help explain this topic.)
The Hardy-Weinberg principle is used to calculate allele and genotype frequencies in a population. It can therefore be used to predict the number of defective individuals in a population. The Hardy-Weinberg equation is

$$p^2 + 2pq + q^2 = 1$$

Consider a pair of alleles Aa; let p = frequency of A, and q = frequency of a.
In the population, $p + q = 1$.
Assuming random mating then in the next generation
AA = p^2; Aa = 2pq; aa = q^2.
Example:
A recessive allele confers resistance to an insecticide in a particular insect species. Explain how the allele is distributed if 36% of the insect population is resistant.

rr = 36% or 0.36 i.e. $q^2 = 0.36$ q = 0.36 = 0.6
Since $p + q = 1$ p = 0.4 $p^2 = 0.16$
2pq = 2 x 0.4 x 0.6 = 0.48 or 48%
Therefore allele distribution is as follows:
RR (p^2) = 0.16 (16%) Rr (2pq) = 0.48 (48%) rr (q^2) = 0.36 (36%)

Sometimes, variations in gene frequencies in populations occurs by chance. This is known as random **genetic drift**. It may be an important evolutionary mechanism in small or isolated populations. Say an allele occurs in 1 % of the members of a species. In a large population, say 1 000 000, then 10 000 individuals may be expected to possess the allele. By chance, the population of individuals with the allele will not be significantly altered in the next generation. If, however, the population is much smaller, say 1000 individuals, only one will carry the allele. By chance, this one may fail to mate and pass on the allele and so it will be lost from the population altogether.
An important case of genetic drift is when a few individuals become isolated from the rest of the species and start a new population, e.g. when a few individuals colonise an isolated island or some new habitat. These founder members of the new population are a small sample of the population from which they originated. By chance, they may have a very different gene

frequency. While the founder population remains small it may undergo genetic drift and become even more different from the large parental population. This process is called the founder effect. The effect undoubtedly contributed to the evolutionary divergence of Darwin's finches after strays from the South American mainland reached the remote Galapagos Islands.

Disasters such as earthquakes, floods, and fires may reduce the size of a population drastically. The result is that the genetic makeup of the surviving population is unlikely to be representative of the makeup of the original population. By chance, certain alleles will be over represented among survivors, other alleles will be underrepresented and some alleles will be eliminated completely. Genetic drift may continue to affect the population for many generations, until the population is again large enough for sampling errors to be insignificant. Because alleles for at least some loci are likely to be lost from the gene pool, the overall genetic variability in the population is usually reduced.

Isolation and speciation

In theory, any individual in a population is capable of breeding with any other. However, breeding sub-units may become separated in some way i.e. isolated, and evolve along separate lines. If, reunited after many generations, they were found to be incapable of breeding successfully with each other, they would have become a separate species. You should understand how separation by geographical features, habitat changes, changes in body form and changes in breeding mechanisms may lead to the formation of new species.

The origins of species

Within a population of one species there are groups of interbreeding individuals. Each of these breeding sub-units is called a **deme**. New species arise when some barrier to reproduction occurs so that the gene pool is divided and different combinations of mutants develop in each pool division. Such a barrier which effectively prevents gene exchange between demes is called an isolating mechanism. If the separation is long-term, eventually the two groups will be so different that two new species incapable of interbreeding are formed. For new species to develop from a population, some form of isolating mechanism is required.

There are two main forms of **isolating mechanisms**:

- **Reproductive isolation** within a population occurs when organisms inhabiting the same area become reproductively isolated into two groups for reasons other than geographical barriers. Species formation occurring in demes in the same geographic area is known as **sympatric speciation**. The barriers to breeding include the following mechanisms:
  - ✓ Behavioural isolation - in animals with elaborate courtship behaviour the steps in the display of one subspecies fails to attract the necessary response in a potential partner of another subspecies.
  - ✓ Mechanical isolation – the genitalia of the two groups may be incompatible.
  - ✓ Gametic isolation – in flowering plants pollination may be prevented because the pollen grain fails to germinate on the stigma whereas in animals sperm may fail to survive in the oviduct of the partner.
  - ✓ Hybrid inviability – despite fertilization taking place development of the embryo may not occur.
  - ✓ Hybrid sterility – when individuals of different species breed, the sets of chromosomes from each parent are different. These sets are unable to pair up during meiosis and so the offspring are unable to produce gametes.
- **Geographical isolation** between populations occurs when the population becomes geographically split into separate demes; the evolution of a new species is very probable, given time. This sort of speciation is known as **allopatric speciation**.

Evolution and selection

Evolution is the process by which new species are formed from pre-existing ones over a period of time. The basis of contemporary thought surrounding the theory of evolution was first put forward by Alfred Wallace and Charles Darwin. In 1859 Darwin proposed natural selection as the force that causes changes in populations. More recently biologists have realised that natural selection can also maintain variation and therefore stabilise a population.

Natural selection

Darwin's observations of variation within a population and the tendency for the adult population to be stable in size led to the development of the idea of natural selection. The theory is based on the following observations:

- In any population there is **variation**.
- Individuals within a population have the potential to produce large numbers of offspring yet the number of adults tends to stay the same from one generation to the next.

From these observations, two deductions were made:

- ✓ There is a struggle for survival (**competition**) with only the 'fittest' surviving.
- ✓ The individuals that survive and **reproduce** pass on to their offspring the characteristics that enable them to succeed (i.e. a **selective advantage**).
- ✓ In time, a group of individuals that once belonged to the same species may give rise to two different groups that are sufficiently distinct to belong to two separate species.

If the environment or conditions change, then the features needed to survive in it will change, so natural selection is a continuous process.

During his voyage on the HMS Beagle, Darwin found evidence of **adaptive radiation** in the **finch** population. He observed fourteen species, differing greatly but particularly in beak morphology, reflecting differences in feeding habits. It was suggested that soon after the Galapagos Islands formed they were populated by a flock of finches from the mainland. These then evolved by natural selection to fill the available vacant niches, as other perhaps better-adapted forms that would have been in competition with the finches had failed to reach the Islands.

❑ ***Action*** List the different types of beaks found in finches.
Describe how the beak shapes are adapted to different feeding niches.

When environmental conditions change, there is a selection pressure on a species causing it to adapt to the new conditions. This pressure determines the spread of an allele within the gene pool.
Darwin proposed natural selection as the force that causes changes within populations. More recently biologists have realised that natural selection can also maintain variation if there is no need to change and therefore stabilise a population.

(The following may be useful as a means of understanding the concept but will not be tested in the exam.)

The types of selection are:

- **Directional selection** is when an environment favours all the individuals carrying one gene at the expense of the individuals carrying an alternative gene.
- **Stabilising selection** results in several different types of individuals being maintained in a population so that genetic variation is maintained.
- **Disruptive selection** favours both extremes of variants and leads to a divergence of the phenotype i.e. to the emergence of two distinct forms. This form of selection is rare but is more likely to result in the formation of a new species.

Human influences as a selection pressure

- You need to study how the use of pesticides has applied directional selection to pest populations resulting in the development of pesticide resistance.
- Humans have selectively bred plants and animals for thousands of years. Traditionally, this would be accomplished by mating together individuals that showed desirable characteristics.

Resistance

'Super pests' have evolved through the use of drugs and pesticides that have eliminated competition allowing a resistant form to thrive.

Antibiotic resistance

Normally bacteria, e.g. *Staphylococcus aureus*, are killed by the antibiotic, penicillin. If a mutation occurs and a bacterium is formed which is resistant to penicillin, it will have a selective advantage, i.e. it will be more likely to survive, and when it reproduces, it will produce offspring like itself. Bacteria reproduce asexually so all the offspring of this mutant individual will be resistant to penicillin.
(This topic has been covered in Unit 4, page 33.)

Pesticide resistance

The changes made to natural populations as a result of man-made changes in the environment has led to an increase in organisms that do great damage and cause economic losses. Various pesticides have been used to kill pest species. Their use has applied directional selection to the pest populations and this has resulted in the development of resistant populations. Where a single pesticide has been widely and continuously used, many populations now have a high proportion of resistant individuals. Recent control methods have therefore used lower concentrations of a particular pesticide and applications of mixed pesticides. This reduces the selection pressure for a particular resistant gene and resistance evolves more slowly.

Resistance in rats

Rats, *Rattus norvegicus,* have become resistant to the anticoagulant 'warfarin' which has been used on a large scale to control their populations. A dominant allele R at a single locus in rats confers resistance. However, this allele also confers a requirement for vitamin K.

- ✓ Heterozygotes (Rr) are resistant to warfarin and have only a small requirement for vitamin K.
- ✓ Homozygotes (RR) are resistant to warfarin but have a massive requirement for vitamin K which is difficult to meet.
- ✓ Homozygotes (rr) are killed by warfarin but have a much better chance of survival than RR rats if warfarin is absent from the environment.

This is an example of **heterozygote advantage** where the heterozygotes are favoured by selection and both alleles R r will be maintained in the population with all three genotypes RR, Rr and rr being produced in each generation. Then in the rat population both warfarin sensitive and warfarin resistant alleles are maintained in areas where warfarin is used as a selective agent.

❑ ***Action*** Explain how the use of antibiotics in animal feed leads to the development of resistance of bacteria in humans.

Artificial selection

Artificial selection (selective breeding) of animals and plants makes use of variations that occur within a population. In this case, humans choose organisms showing desirable characters and breed only from these. This process of artificial selection mimics natural selection and provides evidence that selection can lead to the development of characteristics and the production of very distinct forms of organisms, as seen in many domestic animals and plant species.
Artificial selection:

- ✓ is carried out by humans to obtain plants or animals with the characteristics humans require.
- ✓ may take many years to develop organisms with the required characteristics.
- ✓ produces organisms belonging to the same species, but are often described as different breeds or varieties.

There are two basic methods of artificial selection:

- **inbreeding** – occurs when the gametes of close relatives fuse.

The problem with inbreeding is that it promotes homozygosity i.e. it increases the chance of a harmful recessive gene expressing itself, since there is a greater risk of a double recessive individual occurring e.g. plant species inbred over many generations show a degree of loss of vigour, size and fertility. This is called inbreeding depression. At intervals it is necessary to introduce new genes by outbreeding.

- **outbreeding** – occurs by the crossing of unrelated varieties.

Outbreeding promotes heterozygosity. It introduces hybrid vigour where the organisms sometimes grow more strongly. It arises when the new sets of chromosomes are complementary in their effects.
Occasionally crosses have occurred between plants of different species e.g. the development of modern wheat.

There are five main steps in any **breeding programme**:

1. Look for individuals with the characteristics you require, e.g. resistance to disease.
2. Breed together two of these individuals (or self-pollinate if it is a plant).
3. Collect the offspring, and select those which have the characteristics you require.
4. Breed from these offspring.
5. Repeat these steps over many generations.

An example of selective breeding in animals is milk yield in cattle.
To develop cows with an increased milk yield you would carry out the following steps:

1. test the milk yield of the cows you have to breed from;
2. select the cow with the highest milk yield (A);
3. select a bull descended from a cow with a high milk yield (B);
4. cross A and B, and select female calves;
5. wait for these calves to mature, then test their milk yield;
6. select the cow with the highest milk yield (C), then repeat these steps over several generations.

❑ ***Action*** Give definitions for the following terms:
variety, pure line, breed, inbreeding, outbreeding, hybridisation, hybrid vigour.

Recent new scientific developments, such as embryo transfer and artificial insemination, have introduced improvements to animal breeding techniques. These topics are covered in section 5.5 on page 61.

Biodiversity and causes of extinction

Biodiversity is the variety of species on earth; extinction is the loss of species. Since the beginning of settled agriculture and the development of urban living, human activities have had various deleterious affects on the environment and on the existence of 'naturally' occurring animals and plants.

❑ ***Action*** You may wish to recap information from Unit 2.6, pages 66-70: "Human activities affect the environment"

How farming practice can reduce or enhance biodiversity

The need for increased food production has been satisfied in the following ways:

✓ Greater use of fertilisers and pesticides.
✓ Increased mechanisation.
✓ Increased size of fields.
✓ Improved strains of plant and animal species.

The demands of agriculture often conflict with the need for conservation e.g. farmers remove hedges from their land because:

✓ They take up space.
✓ Easier access is needed for large machinery.
✓ They harbour pests, diseases and weeds.

Hedges have considerable conservation value because:

✓ They prevent soil erosion by the wind.
✓ They provide a habitat for plants and animals.
✓ They provide food for animals which may not live in the hedgerows.

There is therefore a need for the management of farms to ensure sustainability and to reduce the impact on wildlife. This may be achieved by such measures as the control of pesticide use, the maintenance of habitat variety and the prevention of erosion.

Endangered species

The vast majority of Earth's earlier occupants, including the large and once dominant, dinosaurs and tree ferns, have become extinct largely as a result of climatic, geological and biotic changes. At the present time, human activity has taken over as the main cause of species evolution. Many of the larger mammals e.g. mountain gorillas, giant pandas, tigers and polar bears are threatened.
Their decline in numbers have three main causes:

- Loss of habitat;
- Over-hunting by humans;
- Competition from introduced species.

Other species are also threatened by additional causes such as:

- Deforestation;
- Pollution, e.g. PCBs and oil;
- Drainage of wetlands.

It is now recognised that each species may represent an important human asset, a potential source of food, useful chemicals, or disease-resistant genes. There is therefore a need for **species conservation**, the planned preservation of wildlife. Threatened species are now assigned to one of five categories: endangered, vulnerable, rare, indeterminate, out of danger.

Conservation of genetic sources

- ➢ *The conservation of species ensures the conservation of existing gene pools.*
- ➢ *The importance of conserving existing gene pools in the wild and in captivity.*
- ➢ *The ethical reasons and the loss of potentially useful genes to man and the species.*

Present-day plants and animals used in agriculture and horticulture have been developed from plants and animals that were originally in the wild. Breeding increases genetic uniformity with the loss of rarer alleles. In the past breeders may have neglected some important qualities, such as, resistance to cold, disease resistance, etc. These need to be added back into highly cultivated varieties, using the wild plants and animals as a gene bank. If habitats and the wildlife they house are threatened this may no longer be possible. There is also concern about the progressive destruction of the tropical rain forests. Among the many trees and shrubs are some with medicinal properties. The extinction of any plant species before their chemical properties have been investigated could amount to an incalculable loss. In recent years there has been much concern about the loss of gene pools and various legislation has endeavoured to prevent the extinction of endangered species.

The following are some of the steps that have been taken:

- Stocks of seeds of 'traditional' varieties of plants are stored in seed banks.
- The establishment of sperm banks.
- The founding of rare breeds societies to maintain old, less commercial varieties of animals.
- The protection and breeding of endangered species in specialised zoos.
- Reintroduction programmes, e.g. Red Kite in mid Wales.
- Global organisations, such as the World-Wide Fund for Nature, mount continuing campaigns to promote public awareness.
- International co-operation restricting trade, e.g. in ivory and whaling.
- In the UK, the Nature Conservancy Council is the government body that promotes nature conservation and gives advice to government and to all those whose activities affect wildlife and their habitats.
- ✓ It produces a range of publications.
- ✓ It proposes schemes of management for each of the major ecosystem types, endeavouring to conserve species diversity.
- ✓ It establishes nature reserves managed by wardens.

Education and legislation have also played their part in conservation.
Legislation has been introduced to protect endangered species and to prevent over-grazing, over-fishing, hunting of game, collection of birds' eggs, picking of wild flowers, and collection of plants.

❑ ***Action***

1. Consider the economic and ethical reasons for conservation.
2. Write an essay:
   What are 'endangered species'? For what reasons have some species become endangered? How may species be conserved for future generations?

5.3 Sexual reproduction of plants

The flowering plants or angiosperms are the most successful of all terrestrial plants. The flower is the organ of reproduction and is usually hermaphrodite. In angiosperms the female part, the ovule, is never exposed but is enclosed within a modified leaf, the carpel. You will need to study the structure of a typical flower as well as how it produces the male and female gametes, pollen and ovules.

Flower structure

❖ An **insect-pollinated flower** generally has bright colour, patterns and scent to attract potential pollinating insects.
A typical flower is made up of four sets of modified leaves, each set being referred to as a whorl:

- **Calyx** (outer whorl) – **sepals**, which protect the flower in bud.
- **Corolla** – **petals**, brightly coloured to attract insects.
- **Androecium** – **stamens**, which comprise a long **filament** at the end of which are the **anthers** which produce **pollen grains** containing the male gamete. As well as supporting the anther the filament contains vascular tissue which transports food materials necessary for the formation of pollen grains. The anther is usually made up of four **pollen sacs** arranged in two pairs, side by side.
- **Gynoecium** (inner whorl) – made up of one or more **carpels**. Each carpel is a closed structure inside which one or more **ovules** develop. Each ovule develops from the nucellus and becomes surrounded by two integuments. The lower part of the carpel, which surrounds the ovules, is called the **ovary** and bears at its apex a stalk-like structure, the **style**. This ends in a receptive surface, the **stigma**.

❑ ***Action*** Draw a labelled diagram to show the structure of a named regular insect-pollinated flower - Primrose (*Primula*).
Annotate the diagram to show the function of the various parts.

❖ **Wind-pollinated flowers** are dull, unattractive and without scent. The petals are usually absent leaving the anthers and stigmas exposed.

❑ ***Action***
1. Draw a labelled diagram of a wind-pollinated flower – Rye grass (*Lolium perenne*).
2. In table form compare a wind- and insect-pollinated flower.

Pollen grain development

- In the early stages of development, each pollen sac contains a mass of diploid **pollen mother cells** surrounded by a nutritive layer called the **tapetum**.
- Each microspore mother cell undergoes **meiosis** to give a tetrad (four) of haploid cells called **microspores**. Each of these cells develops into a **pollen grain.** A thick outer wall, the **exine**, and a thin inner wall, the **intine**, are secreted.
- Inside the pollen grain the haploid nucleus undergoes mitosis to produce **a generative nucleus** and a **pollen tube nucleus**. The generative nucleus will later give rise to the two **male nuclei.**
- As the pollen grains mature, changes take place in the anthers. The cells of the tapetum shrink and degenerate, and fibrous layers develop beneath the epidermis.

- When the pollen is ripe, the outer layers of the anthers dry out and tensions are set up in lateral grooves. Eventually dehiscence occurs and the edges of the pollen sacs curl away exposing the pollen grains.

❑ ***Action***
1. Draw an annotated diagram of T.S. anther.
2. Draw a labelled diagram of a pollen grain, showing the generative nucleus and pollen tube nucleus.

Ovule development

- In the early stages of the development of an **ovule,** the nucellus arises as a tiny lump of tissue growing out from the placenta of the carpel. At is apex a **megaspore mother cell** develops and undergoes **meiosis** to form **four haploid megaspores**.
- Normally only **one** of these cells continues to develop, the rest abort. The single megaspore gets bigger, gaining its nutrition from the nucellus tissue.
- It undergoes **three successive mitotic divisions** and forms an **embryo sac** containing **eight haploid nuclei.**
- In the embryo sac the eight nuclei are arranged in a definite pattern, with initially a group of four haploid nuclei at each pole.

✓ One nucleus from each polar group moves to the centre of the embryo sac.
These are the **polar nuclei**.

✓ Three cells are situated at the micropyle end of the ovule, making up what is referred to as the egg apparatus, the central cell being the female gamete or **egg cell**.

✓ At the opposite end is another group of three cells called the antipodal cells. These play no further part in the process.

❑ ***Action*** Draw a diagram to show a mature ovule and label the following structures; integuments, polar nuclei, egg nucleus and micropyle.

Pollination and fertilization

A key feature of the success of flowering plants is their relationship with animals. Pollen grains have no power of independent movement and have to be transferred to the female part of the flower to ensure fertilisation. Flowering plants have evolved the strategy of attracting animals, particularly insects, to their flowers, feeding them and exploiting their mobility to transfer pollen from flower to flower. Some plants are pollinated by the action of wind.

Pollination

Pollination is necessary in order that the pollen grains, containing the male gametes, are brought into contact with the gynoecium so that fertilisation can be achieved. This means that pollen grains must be transferred from the ripe anther to the receptive stigma.

Self-pollination

In some species self-pollination occurs and the pollen from the anthers of a flower need only be transferred to the stigma of the same flower or another flower on the same plant.

Cross-pollination

In a large number of species, cross-pollination is more normal, where pollen is transferred from the anthers of one flower to the stigma of another flower on another plant of the same species. There are two main methods of pollination:

- by wind
- by animal vectors such as insects, birds or mammals

A comparison of self and cross pollination

The two forms of pollination have very different genetic consequences.

- Self-pollination leads to self-fertilization, cross-pollination to cross-fertilization.
- Self-fertilized species depend on random assortment and crossing over during meiosis, and on mutation to bring about variation in the genomes of male and female gametes.
- Self-fertilized species therefore display less genetic variation than cross-fertilized species that are produced from gametes from two different individuals.
- There are advantages to inbreeding because it can preserve good genomes which may be suited to a relatively stable environment. Outbreeding is of greater evolutionary significance because in the struggle for survival some genomes are more successful than others.

Many flowering plant species have evolved a variety of mechanisms to ensure that cross-pollination takes place, so ensuring outbreeding. Anthers and stigma may mature at different times, they may be at different levels in the flower or there may be separate male and female flowers on different plants.
For example, the primrose is a dimorphic plant in which there are two types of flower, pin-eyed and thrum-eyed. They differ in the length of style as well as in the size and chemical composition of their pollen.

❑ ***Action*** Draw labelled diagrams of the two forms of primrose.

In the pin-eyed flower the stigma is above the stamens so preventing self-pollination.
In the thrum-eyed flower if pollen falls onto their own stigmas the pollen will not germinate. So even though self-pollination occurs in the thrum-eyed flower there is a mechanism (incompatibility) to prevent this leading to successful fertilization of the ovule and the production of seed.

Fertilisation

In flowering plants the female gamete is protected within the carpel and the male gamete can only reach it via the pollen tube.

- On landing on the stigma the pollen grains absorb water and germinate producing a pollen tube.
- The pollen tube grows down the style controlled by the tube nucleus, followed by the generative nucleus which divides by mitosis to give two male nuclei.
- The pollen tube enters the embryosac via the micropyle.
- As the pollen tube penetrates the embryo sac, the tip opens and the two male gamete nuclei enter, the pollen tube nucleus having disintegrated earlier.
- The first **male nucleus** fuses with the female, or **egg nucleus**, to form a **zygote**.
- The second **male nucleus** fuses with the **two polar nuclei** to form a **triploid endosperm nucleus**.
- These events constitute a double fertilisation. This is an unique feature of the flowering plants.

❑ ***Action*** Draw a diagram of an ovary containing a single ovule showing a pollen tube as it enters through the micropyle. Use matching colours to show which nuclei fuse together.

Development of the seed

Double fertilisation results in the formation of the food storing endosperm and embryo.
Some seeds retain the endosperm some absorb it into the cotyledon(s) e.g. broad bean.
The seed develops from the fertilised ovule and contains an embryonic plant and a food store.
After a period of dormancy and when environmental factors are favourable stored food will be mobilised and the seed will germinate.

- The diploid zygote nucleus begins to divide by **mitosis** and eventually an embryo, consisting of a **plumule**, a radicle and one or two **cotyledons**, is formed.
- The endosperm nucleus also divides mitotically forming a large number of nuclei and eventually an **endosperm** tissue develops. These cells become filled with stored food materials, providing reserves for the developing embryo.
- The **integuments** fuse together and become the seed coat or **testa**.
- The **carpel wall**, the pericarp, becomes the **fruit.**
- The **ovule** becomes the **seed**.

Structure of the seed

❑ ***Action*** Draw a labelled diagram of broad bean (*Vicia faba*) – internal and external. Label – micropyle, testa, position of radicle, plumule, cotyledons.

Germination

Germination of the seed usually occurs during a period of dormancy. The three main requirements for successful germination are:

- a suitable **temperature** – this is related to enzyme activity;
- **water** – needed for the mobilisation of enzymes, vacuolation of cells and for transport;
- **oxygen** – energy is needed for growth, production of ATP.

Mobilisation of food reserves during germination

Food reserves in seeds are insoluble in water and cannot be transported in the seedling. They must be broken down into relatively simple soluble substances which dissolve in water to be moved the growing apices of the plumule and radicle. Water is taken up rapidly by the seed in the initial stages, causing the tissues to swell as well as mobilising the enzymes. The testa, or seed coat, will rupture as the radicle pushes its way through. The radicle will grow downwards and the plumule upwards. The enzyme, amylase, hydrolyses starch into maltose which is transported to growing points.

5.4 Human reproduction

In the human reproductive system the gametes are produced in special paired glands called gonads. The male gametes or sperm are produced in the testes and the female gametes or egg cells in the ovaries. You need to study the development of gametes which is known as gametogenesis.

Male reproductive system

The male system consists of a pair of **testes**, contained in an external sac, the **scrotum**. The testes produce spermatozoa; the **penis**, which is an intromittent organ; genital ducts connecting the two; and various accessory glands which provide constituents for the semen. Each testis consists of about a thousand **seminiferous tubules**, which produce the spermatozoa. The seminiferous tubules also contain interstitial cells that produce the male hormone, testosterone. When sperm have been produced they collect in the vasa efferentia and then pass to the head of the **epididymis**, where they mature. They then pass along the coiled tube to the base of the epididymis, where they are stored for a short time before passing via the **vas deferens** to the urethra during ejaculation. Before it joins the urethra it combines with the duct leading from the **seminal vesicle**, forming the ejaculatory duct. The seminal vesicles produce a mucus secretion which helps the sperms mobility. The ejaculatory duct then passes through the **prostate gland** which produces an alkaline secretion that neutralises the acidity of any urine in the urethra as well as aiding sperm mobility.

❑ ***Action***
1. Draw a labelled diagram of the male reproductive system.
   Annotate the diagram with the functions of the structures above shown in bold.
2. Draw T.S. testis.

Female reproductive system

There are two **ovaries** each of which produces ova, which are produced in the germinal epithelium. Once formed, the ova develop in follicles. Mature follicles migrate back to the surface when their development is complete so that the ova can be shed.
Ova are passed to the **fallopian tube** (oviduct) which conveys them to the **uterus** (womb). The uterus has muscular walls and is lined internally by a mucus membrane called the **endometrium**. The uterus opens into the **vagina** through a ring of muscle, the cervix.

❑ ***Action*** Draw a labelled diagram of the female reproductive system.
Annotate the diagram with the functions of the structures above shown in bold.

Gametogenesis

The cells of the germinal epithelium of both the testis and the ovary undergo a sequence of mitotic and meiotic divisions to form gametes.
First there is a multiplication stage which involves repeated mitotic divisions to produce **spermatogonia** and **oogonia**. Once formed they grow to full size and then undergo maturation which involve meiotic division, and then differentiation into the mature gametes.

Formation of spermatozoa

These are produced in the germinal epithelium of the seminiferous tubule and are nourished and protected by the **Sertoli cells**. They form **primary spermatocytes** which then form haploid **secondary spermatocytes** after the first meiotic division. After the second meiotic division they form **spermatids** which differentiate into **spermatozoa**.

❑ ***Action*** Draw a labelled diagram to illustrate spermatogenesis.

Formation of ova

Oogonia, which are formed before birth, undergo mitosis to form **primary oocytes,** which then mature just before ovulation. About two million of these are formed in the fetal ovary but only about 450 will later develop into **secondary oocytes**. Before ovulation, a primary oocyte undergoes the first meiotic division to form the haploid secondary oocyte and its first polar body. The secondary oocyte begins the secondary meiotic division, but this is arrested as metaphase unless fertilisation takes place. On fertilisation this division is completed to form a large **ovum** and a second polar body. Once this division has taken place the nucleus of the ovum fuses with that of the sperm to form a **zygote**, which will then develop into an **embryo**.

❑ ***Action***
1. Draw a labelled diagram to illustrate oogenesis.
2. Draw a labelled diagram of an ovary, showing the stages in development of a follicle.

Sexual intercourse

The penis of a sexually excited male becomes erect as blood under pressure fills the spongy 'erectile tissue'. The vagina of a sexually excited female widens and becomes lubricated by the secretion of mucus. During sexual intercourse the penis is inserted into the vagina. Movements of the penis result in the ejaculation of semen into the vagina. The force of ejaculation is sufficient to propel some sperm through the cervix into the uterus, with the remainder being deposited at the top of the vagina. The sperm swim through the uterus into the oviducts by the lashing movements of their tails. However, only a small proportion (i.e. several hundreds) of the 400 million sperm actually reach the site of fertilization in the oviduct (and of course only one actually fertilizes it).

The egg or ovum released from the Graafian follicle of the ovary dies within 24 hours unless fertilized.

Fertilization

Internal fertilisation ensures that the sperm are deposited in the female's reproductive tract. When a sperm fuses with an egg cell a zygote is formed and this develops by mitosis to form a mass of cells. The zygote is implanted in the uterus wall where it continues to develop. You need to study the placenta and describe its role in the development of the embryo until birth.

Sperm are highly fertile for 12 to 24 hours after release into the female tract but can fertilise an ovum only after a process called **capacitation,** which takes several hours.

It involves changes in the membrane covering the **acrosome**, a thin cap over the nucleus. When the sperm reach an oocyte, contact with the jelly coat results in the acrosome membrane rupturing and enzymes are released. The enzymes soften the layers of cells (corona radiata and zona pellucida) surrounding the oocyte. Inversion of the acrosome results in a fine needle-like filament developing at the tip of the sperm and this pierces the already softened portion of the membrane. The whole process is called the **acrosome reaction** and it enables the sperm to penetrate the egg. This entry stimulates reactions of the oocyte that brings about the formation of the **fertilisation membrane**, preventing the entry of further sperm. Entry of the sperm also stimulates the completion of the second meiotic division of the oocyte nucleus. The nuclei of the ovum and sperm are drawn together and fuse to form a diploid nucleus. The fertilised ovum is known as the zygote and it begins to divide by mitosis to form the embryo.

❑ ***Action***
1. Draw a labelled diagram of a mature sperm cell.
2. Describe the events which occur during fertilisation, including digestion of corona radiata and zona pellucida by acrosomal enzymes, entry of sperm head into secondary oocyte, change in zona pellucida preventing entry of further sperm, second meiotic division to produce female gamete, fusion of nuclei to form zygote.

Implantation

As soon as the ovum is fertilised it begins to divide (cleavage) by mitosis until a hollow ball of cells, the **blastocyst**, is produced. The development of the zygote continues during its passage down the fallopian tube. After reaching the uterus the blastocyst embeds in the endometrium. This is called implantation.
(Details of the formation of the extra-embryonic membranes are not required.)

The placenta

The placenta develops in the region where implantation occurred. It develops villi which grow into the surrounding uterine tissue from which they absorb nutrients. These villi form part of the placenta which is connected to the fetus by the umbilical cord. A membrane develops around the fetus and encloses the amniotic fluid which protects the fetus by cushioning it from physical damage.

❑ ***Action*** Draw an annotated diagram of the placenta.

Functions of the placenta

- To nourish the developing fetus. In the placenta the maternal blood is brought close to that of the fetus but they do not mix. Finger-like projections of the membranes grow into the endometrium and become bathed by maternal blood. There is a minimal barrier over a huge surface area allowing the diffusion and active transport of materials across the placenta. Water, amino acids, glucose, lipids, mineral salts, simple proteins, vitamins, hormones, antibodies and oxygen pass from the mother to the fetus.
- The removal of wastes – carbon dioxide, urea and other nitrogenous wastes are transferred to the maternal blood to allow their excretion by the mother.
- To act as a depressurising unit – it protects the fetal circulation from the high maternal blood pressure.
- It acts as a filter – the placenta prevents the passage of maternal hormones that could affect fetal development, while allowing some hormones across it. It also prevents the passage of some pathogens although some viruses and bacteria can cross.
- As the pregnancy progresses the placenta increasingly takes over the role of hormone production.
  It produces:

✓ progesterone which prevents ovulation and menstruation;
✓ human choriogonadotrophin (HCG), which maintains the corpus luteum during pregnancy.

Birth

During pregnancy the quantity of different hormones produced changes, with the amount of oestrogen increasing and that of progesterone decreasing. This triggers the birth process and the hormone **oxytocin** is produced from the posterior lobe of the pituitary gland. This hormone causes contractions of the uterus, which results in birth of the baby, followed a little later by the placenta, or afterbirth.

Lactation

After the birth the anterior lobe of the pituitary gland produces the hormone, **prolactin**, which causes the mammary glands to produce milk.

Hormonal control of reproduction

In all mammals female sexual activity is cyclical. A cycle in the ovary leads to the production of mature ova and a uterine cycle prepares the uterus to receive fertilised eggs. In humans there is a regular 28 day cycle controlled by hormones secreted by the pituitary gland and the ovary.

Menstrual cycle

The events that occur in the course of the menstrual cycle follow a set pattern which is regulated by hormones from the pituitary glands and ovaries. If fertilisation occurs, the normal cyclical pattern is interrupted and an additional source of hormones, the placenta, comes into play. The main hormones involved and their functions are as follows:

- **Follicle stimulating hormone** (FSH), secreted by the anterior pituitary.
  - ✓ causes Graafian follicles to develop in the ovary
  - ✓ stimulates the tissues of the ovary to produce oestrogen
- **Oestrogen**, produced by the ovary
  - ✓ inhibits the production and release of FSH
  - ✓ repairs the uterine wall following menstruation
  - ✓ builds up in concentration during the first two weeks of the menstrual cycle until it stimulates the pituitary gland to produce luteinising hormone
- **Luteinising hormone** (LH), secreted by the anterior pituitary
  - ✓ brings about ovulation
  - ✓ causes a Graafian follicle to develop into a corpus luteum which produces oestrogen and progesterone
- **Progesterone and oestrogen**
  - ✓ inhibit FSH and LH production and therefore stops further follicle development
  - ✓ progesterone only causes development of the uterine wall prior to implantation

❖ If fertilisation does not take place:
the falling levels of FSH and LH, allow the corpus luteum to degenerate and now the concentrations of progesterone and oestrogen also fall. As a result FSH and LH production is no longer inhibited. The thickened epithelium of the endometrium is shed, producing the menstrual flow.

❖ If fertilisation occurs:
progesterone and oestrogen,produced initially by the **corpus luteum** and later by the placenta, are present in high concentration. LH production and then ovulation are inhibited; follicle development and production of oestrogen cease and the uterus lining is maintained.

❑ ***Action***
1. Draw graphs showing changes in the levels of hormones involved in the menstrual cycle.
2. The presence of the hormone HCG in the urine of pregnant women is the basis of most pregnancy tests. Find out how immobilised antibodies are used in pregnancy testing kits. (Link with Unit 4.8.)

There is no sexual cycle **in the male** but **gonadotrophic** hormones are produced from the anterior lobe of the pituitary gland.

- **FSH** stimulates sperm development
- **LH** stimulates the production of testosterone
- **Testosterone** controls the development of male reproductive organs and secondary sexual characteristics.

5.5 Applications of reproduction and genetics

In the past, conventional breeding techniques have been used to improve farm animals, crop and ornamental plants. However, the selection and cross breeding is laborious, time consuming and sometimes unpredictable. Micropropagation provides a rapid method for obtaining large numbers of genetically identical plants. In the future the quality of farm animals may be improved by laboratory based breeding techniques involving embryos.

Cloning

- A clone is an asexually reproduced, hence genetically identical line of cells or organisms.
- Cloning can occur naturally, e.g. bacteria, yeast;plants grown from suckers, bulbs, corms.
- Artificial clones can be formed in animals by separating embryos at an early stage.
- Artificial clones can be formed in plants by taking cuttings and micropropagation.

❑ ***Action*** List the differences between asexual and sexual reproduction.

Cloning of plants - micropropagation

Asexual reproduction involves only one organism and the individuals produced are genetically identical i.e. belong to a **clone**. The technique of micropropagation is based on the ability of the differentiated plant cell to give rise to all the different cells of the adult plant i.e. plant cells are **totipotent**. Micropropagation is sometimes referred to as a test-tube plant culture. It is an extremely cost effective way of producing large numbers of genetically identical plants, which are clones of a single parent.
At the tip of the roots and shoots of plants are areas called meristems. Here the cells divide rapidly by mitosis, so the meristems are known as growing points. If cells are removed from the meristem and placed in suitable conditions, new, genetically identical, plants will develop.

These steps are generally followed in the micropropagation:
1. Use a sharp knife to remove the meristem from the shoot.
2. Cut the meristem into small pieces(explants).
3. Place the explants onto sterile, aerated nutrient medium, e.g. agar jelly.
4. The cells divide by mitosis to produce a callus, which is a mass of undifferentiated cells.
5. The callus is subdivided. and each piece differentiates into a plantlet.
6. When they have reached a suitable size the plantlets are transplanted into sterile soil.

❑ ***Action*** Rewrite the above steps in the form of a flowchart.

❖ The **advantages** of micropropagation are:
- Large numbers of plants can be grown in sterile controlled conditions ensuring a greater survival rate than would be the case if seeds were planted outside.
- Uniformity of crop.
- Cold storage of large numbers of plants in a small space with reduced heating and lighting costs.
- Perpetuation of unique genotypes.
- Reduced space required for transport.
- Pathogen-free when planted out.

❖ The **disadvantages** of micropropagation are:

- Requires aseptic conditions otherwise bacterial or fungal contamination of the culture medium results with subsequent loss of plants.
- The plants are genetically unstable, with an increased rate of mutation in medium-grown cells leading to abnormality in the plantlets. Regular inspection is needed to remove any defective individuals, thus labour costs are higher than with traditional propagation methods.

Cloning of animals

Nuclear transplants from somatic into egg cells

The cloning of animals may involve the transfer of a nucleus from one individual to an egg from which the nucleus has been removed and the subsequent development of this embryo in a host or surrogate.

Embryo manipulation

Clones can be developed by taking an embryo at an early stage of development, e.g. after a test-tube fertilization procedure, and splitting the embryo into separate cells. Each of these cells will then develop into a new embryo, genetically identical to the original.
This technique of embryo surgery has enabled farmers to increase their stock. The fertilised egg is bisected at the 2-celled stage and each half is transplanted into different regions of the uterus, allowing twin births to be the rule, not the exception. This technique is also used to conserve rare breeds, where young embryos of young animals are bisected and successfully transplanted into a surrogate mother of a common breed to produce a new individual of the rare type.

Embryo cloning

This technique allows many genetically identical copies of an animal to be produced. If a mutation occurred in a cow making her a high milk producer, significantly better than other members of the herd, cross breeding her with a bull would reshuffle her genes with a consequent loss of her unique characteristic. Cloning is the only technique that will conserve her unique features for future generations.

❑ ***Action*** Draw flowcharts to illustrate the cloning of animals.

❖ The **advantages** of cloning in animals:

- Cell culture is useful for production of cells in quantity e.g. cancer cells for medical research, monoclonal antibodies;
- The production of a single, identical, genetic line of cells with desirable characteristic may be used to maintain genetic stocks.

❖ The **disadvantages** of cloning in animals:

- In mammals the technique is very expensive and unreliable.
- There may be the inadvertent selection of disadvantageous alleles.
- Progeny may show long term/unforeseen effects such as premature ageing.

❑ ***Action*** Consider the ethical implications of the above applications in plants and animals.

5.6 Species are classified into groups

Until recently it was common to include all living organisms into two kingdoms, plants and animals. Because there are some organisms which can be classified in both kingdoms and some which do not fit into either, the five kingdom classification was proposed. You need to know about the distinguishing features of the five kingdoms.

Living organisms are divided into five kingdoms:
Prokaryotae, Protoctista, Fungi, Plantae, and Animalia.

Prokaryotae

The prokaryotae are unicellular organisms and comprise:

- Bacteria
- Blue-green or Cyanobacteria

They have no internal cell membranes; there is no nuclear membrane; no endoplasmic reticulum, no mitochondria and no Golgi body. They possess a cell wall but it is not made of cellulose.

Protoctista

The members of this kingdom are mostly small eukaryotic organisms, with membrane-bound organelles and a nucleus with a nuclear membrane. In this kingdom are found the organisms which are neither plants nor animals nor fungi. So the kingdom includes algae, the water moulds, the slime moulds and the protozoa.

Fungi

The members of this kingdom are eukaryotic, the body consisting of a network of threads called hyphae, forming a mycelium. There is a rigid cell wall made of chitin. There are no photosynthetic pigments present and feeding is heterotrophic; all members of the group being either saprophytic or parasitic. In some sub groups, the hyphae have no cross-walls, but in others cross-walls, or septa, are present. Reproduction is by spores which lack flagella.

Plantae

The members of this kingdom are multicellular and photosynthetic. The cells are eukaryotic, have cellulose walls, vacuoles containing cell sap and chloroplasts containing photosynthetic pigments.
(You are not required to learn the names of the following sub-groups.)
Some of the phyla of the plantae include:

- Bryophyta – mosses and liverworts
- Filicinophyta – ferns
- Coniferophyta – the conifers
- Angiospermophyta – the flowering plants

The flowering plants are the predominant plant group. They include all our major plant foods. Their flowers have seeds which are enclosed in a fruit formed from the ovary wall.

Animalia

The members of this kingdom are multicellular, heterotrophic, eukaryotes, cells lacking a cell wall, exhibit nervous coordination.
(You are not required to learn the names of the following sub-groups.)
This group includes the following phyla:

- Cnidaria (jellyfish)
- Platyhelminthes (flatworms)
- Nematoda (roundworms)
- Annelida (segmented worms)
- Mollusca (molluscs)
- Arthropoda (arthropods)
- Echinoderms (starfish)
- Chordata.

All but the simplest of the chordates have a vertebral column and are therefore referred to as vertebrates. The vertebrate chordates are subdivided into the following classes:

- fish
- amphibians
- reptiles
- birds
- mammals.

Principles of taxonomy

The sorting of living organisms into groups of a manageable size is known as taxonomy or classification. You should know about the principles of modern classification which shows how organisms may be related through evolution by the number of features they share.

Grouping organisms in a hierarchical scheme

The natural classification was devised by Linnaeus in the 18thCentury. In this scheme organisms are grouped together according to their basic similarities. Such homologous characters are those which have a similar origin, structure and position. A hierarchical system has been devised to distinguish large groups of organisms with a series of rank names to identify the different levels within the hierarchy.

- A **taxon** is a level in this classification hierarchy and is an assemblage of organisms sharing some basic features.
- A **species** is a group of organisms which share a large number of common characteristics and which can interbreed to produce fertile offspring. (Fertile offspring are only produced when homologous chromosomes can pair at meiosis.)
- Similar species are grouped together into a **genus**, similar genera into family, families into **order**, orders into **class**, and classes into **phyla**. Phyla are subdivisions of **kingdom**.

❑ ***Action*** Write out the scheme of taxa within the classification system in order of size, starting with kingdom and ending with species.
As one progresses up the hierarchy do the similarities between organisms increase or decrease?

Binomial system

Organisms are named according to the binomial system introduced by Linnaeus in 1753. Each is named by its genus and species. Latin is used as an international language so that an organism is given precise identification worldwide, whereas the common name is not e.g. the tiger belongs to the genus *Panthera* and the species is *tigris*, so its name, written in full, is *Panthera tigris.*
The following describes the classification of the tiger:
You need to be able to classify the tiger, with reasons for being placed in each taxon.

Taxon		Reasons for inclusion in taxon / feature
kingdom	Animalia	heterotrophic, multicellular, no cell walls
phylum	Chordata	notochord, post anal tail
class	Mammalia	hair, mammary glands, internal fertilisation
order	Carnivora	powerful jaws with vertical movement only, large canine teeth, carnassial teeth
family	Felidae	shortened muzzle, fewer teeth than Canidae, retractile claws
genus	Panthera	large size (over 3 metres long)
species	*P. tigris*	yellow / tawny fur with dark stripes, larger than lion and more solitary

5.7 Control systems regulate life processes

Homeostasis

The term homeostasis is used to describe the mechanisms by which a constant internal environment is achieved. An example of the advantage of such constancy is the control of water content and solute composition of body fluids. This is known as osmoregulation. You need to know how the kidney operates not only as an organ of removal of nitrogenous waste but also as an organ of osmoregulation.

The homeostatic process

- All homeostatic processes involve a **detector** (or sensor) which monitors the factor being controlled.
- When the detector senses a change from the normal, it informs a **controller,** a coordinating system which triggers an appropriate method of correcting the deviation.
- The controller communicates with one or more **effectors** which carry out the corrective procedures.
- Once the correction is made and the factor returned to normal, information is fed back to the detector which then 'switches off'.
- This is what happens in most biological control systems, i.e. the controller is no longer alerted to the deviation from the normal. This is called a **negative feedback**.

❑ ***Action*** Construct a flowchart to illustrate the homeostatic process.

The kidney

The kidney has two functions:

- Removal of nitrogenous metabolic waste from the body.
- Osmoregulation, the mechanism by which the balance of water and dissolved solutes is regulated.

The production of urea

Urea is a poisonous chemical made by the liver. If there is too much protein in the diet any excess has to be broken down (as it cannot be stored like carbohydrates and fats). The amino acids, which make up protein, are **deaminated** in the liver. The reaction produces ammonia, which is quickly converted to urea. Urea is released into the blood, and travels around the body until it is removed by the kidneys.

Structure of the kidney

❑ ***Action***
1. Draw a labelled diagram of the mammalian urinary system.
   Label - renal artery, renal vein, kidney, ureter, bladder, urethra.
2. Draw a section through a kidney (L.S.)
   Label – cortex, medulla, pelvis; include the position of a nephron.

Kidneys are made up of about a million **nephrons**. Each nephron is made up of four functional parts –
a **Bowman's capsule**, a **proximal tubule**, the **loop of Henle** and a **distal tubule** leading to a **collecting duct**.

Within each Bowman's capsule is a knot of blood capillaries known as the **glomerulus.** The blood supply to the nephron begins as an afferent arteriole serving the glomerulus. From the glomerulus the blood is carried by the efferent arteriole to two other capillary structures:

- a capillary network serving the proximal and distal convoluted tubules.
- a capillary network running beside the loop of Henle and known as the **vasa recta**.

❑ ***Action*** Draw a single nephron.
Label - afferent arteriole, efferent arteriole, glomerulus, Bowman's capsule, proximal convoluted tubule, loop of Henle, distal convoluted tubule, collecting duct.

Functioning of the kidney

The functioning of the nephron involves three different processes, ultra filtration, selective reabsorption and secretion. The osmoregulatory function of the kidney is controlled by hormones.

Ultra filtration

- Ultrafiltration is filtration under pressure that separates small soluble molecules from the plasma.
  It is the process by which small molecules such as water, glucose, urea and salts are filtered from the knot of capillaries, the glomerulus into Bowman's capsule.
- The structure of the glomerulus and capsule allows ultrafiltration. The basement membrane of the capillary forms the selective barrier between the blood and the nephron and it acts as a molecular sieve.
- The sieve action allows smaller-sized molecules to pass through, but retains in the capillaries the blood proteins and cells.
- Most of the pressure producing the filtration comes from the hydrostatic pressure of the blood in the glomerular capillaries. This pressure is amplified by the pressure in the capsule produced by the narrow efferent vessels and also by the water potential in the blood produced by the colloidal plasma proteins. The glomerular pressure can be altered by changes in the diameter of the afferent and efferent arterioles entering and leaving the glomerulus.

❑ ***Action*** Draw a diagram of the detailed structure of the site of ultrafiltration.
(Include basement membrane, endothelial cells of blood capillary, podocytes.)

Selective reabsorption

- All the glucose and most of the water and salt (sodium and chloride ions) are reabsorbed in the proximal convoluted tubule. Small amounts of water and salt are reabsorbed in the distal tubule. Most of the remaining water is reabsorbed in the collecting duct by a mechanism which involves the loop of Henle.
- Reabsorption of glucose and salts takes place by active transport. Water is reabsorbed passively by osmosis following the transport of salt.
- The loops of Henle collectively concentrate salts in the tissue fluid of the medulla of the kidney.
- The high concentration of salt then causes an osmotic flow of water out of the collecting ducts, thereby concentrating the urine and making it hypertonic to the blood.
- The loop of Henle uses the principle of a **hair-pin counter current multiplier.**
- As fluid flows up the ascending limb, Na^+ and Cl^- is actively pumped out of the ascending limb into the tissue fluid of the surrounding medulla where a low water potential is created.

The ascending limb is relatively impermeable to water while the descending limb is permeable. Water leaves the filtrate of the descending limb by osmosis and is carried away by the blood in the vasa recta.

- The contents of the descending limb become progressively more concentrated and reaches its maximum concentration at the tip of the loop; as it flows up the ascending limb the fluid becomes more and more dilute. Since the surrounding fluid also becomes more concentrated an osmotic gradient is maintained down to the tip of the loop. The effect at one level is slight, but the overall effect is multiplied by the length of the hair-pin. The result is that a region of particularly high salt concentration is produced in a deep part of the medulla.

❑ ***Action***
1. Draw detail of cells from the wall of the proximal convoluted tubule. Describe how the cells are ideally adapted for reabsorption, using the terms –large surface area, due to microvilli and basal channels, numerous mitochondria, closeness of blood capillaries etc.
2. Suggest how the loop of Henle is modified in desert animals.

Secretion

Secretion of further substances not required by the body may take place in the distal convoluted tubule e.g. hydrogen and hydrogencarbonate ions. This is important in the control of **plasma pH**, which must be maintained at 7.4. If the plasma pH falls, hydrogen ions are excreted by the kidney; if the plasma pH rises, hydrogencarbonate ions are secreted.

Role of antidiuretic hormone (ADH) in osmoregulation

- The permeability of the walls of the collecting duct, like the walls of the distal convoluted tubule, is subject to hormonal control. This hormonal effect, together with the hypertonic interstitial fluids built up by the loop of Henle in the medulla, determine whether hypotonic or hypertonic urine is released from the kidney.
- ADH makes the walls of the collecting duct permeable so that water is reabsorbed and the urine has a concentration close to the concentration of the tissues near the bottom of the loop i.e. hypertonic to the general body fluids.
- The amount of water reabsorbed is controlled by a **feedback system**. Negative feedback restores the normal osmotic concentration if blood is diluted or becomes more concentrated.
- A fall in water potential of the blood may be caused by one or a combination of factors – reduced water intake, sweating, intake of large amounts of salt. The concentration of sodium chloride in the blood is an indirect indication of the volume of water in the body. The hypothalamus is sensitive to the concentration of sodium chloride in the blood flowing through it. If the water content is low a fall in water potential is detected by **osmoreceptors** (osmotic receptors) in the hypothalamus and results in nerve impulses passing to the posterior pituitary gland which then releases **ADH**. The hormone increases the permeability of the distal convoluted tubule and the collecting duct to water and so more water is reabsorbed into the blood. Consequently the small volume of urine eventually eliminated is relatively concentrated.
- The parts of the osmoregulatory system which regulate water content are:

✓ hypothalamus - detector
✓ posterior lobe of pituitary (ADH) - coordinator
✓ collecting ducts of kidney - effector.

❑ ***Action*** Describe how negative feedback restores the normal concentration if the water potential of the blood rises, i.e. becomes less negative.

5.8 Sensitivity

For the nervous system to carry out its function effectively it is dependent upon a continuous input of information from inside the body and from the environment. This input is initiated by sensory **receptors**, which range from specialised sensory cells to the most complex sense organs such as the mammalian ear. You should study the structure and function of the ear as an organ for receiving, amplifying and transducing sound waves into electrical impulses for interpretation by the brain.

Mode of action of receptors

Responses to all stimuli involve the reception of information and its transfer from the receptor to an effector via the nervous system.
Sensory receptors are detecting one form of energy and converting it into electrical energy; i.e. nerve impulses. They are acting as **transducers**. Sensory cells at rest have a polarised membrane like nerve cells. When stimulated, the membrane depolarises, producing a generator potential. The magnitude of the generator potential varies with the strength of the stimulus. When this reaches the threshold, an action potential is produced in the sensory axon leaving the sensory cell. As in a neurone the mechanism involves the exchange of sodium and potassium ions.

The ear

Sensory cells are often found in groups, usually together with other tissues, forming sense organs.
(*You are required to study the ear as an organ of hearing. Details of its function as an organ of balance is **not** required.)*
The ear can be divided into three regions:
- ✓ the outer ear(air-filled)
- ✓ middle ear (air-filled)
- ✓ inner ear (fluid-filled).

❑ ***Action*** Draw a diagram of the human ear.
(Labels - pinna, external auditory meatus, ear drum, malleus, incus, stapes, oval and round window, cochlea, auditory nerve.)

Structure of the cochlea

The **cochlea** consists of a coiled tube which is divided longitudinally into three parallel canals (i.e. one canal above the other in a row). The three canals are separated from each other by membranes:
- ✓ **Reissner's membrane** between the upper and middle canals;
- ✓ **basilar membrane** between the middle and lower canals.

All three canals are filled with fluid:
- ✓ **endolymph** in the middle canal;
- ✓ **perilymph** in the upper and lower canals.

❑ ***Action*** Draw a T.S. through the cochlea.
(Labels - endolymph, perilymph, tectorial membrane, Reissners membrane, sensory hairs.)

Into the middle canal projects the **tectorial membrane**, which runs parallel with the basilar membrane for the whole length of the cochlea. Receptor cells span the gap between the basilar and tectorial membranes. They have their bases in the basilar membrane where they are connected to nerve fibres that join the **auditory nerve**. At the other end the receptor cells have fine **sensory hairs** which just reach the tectorial membrane.

Functions of the parts of the ear

- The **pinna** channels the sound waves towards the **ear drum**.
- The sound waves travel to the ear drum which vibrates.
- The movements of the ear drum are transmitted by the three ear ossicles - **malleus**, **incus**, **stapes** (which also amplify the vibrations) to the **oval window**.
- Movement of the oval window causes displacement of the fluid in the upper canal of the cochlea of the inner ear, which in turn causes movement of the **Reissner's membrane**.
- This displaces fluid in the middle canal which moves the **basilar membrane**, thereby displacing fluid in the lower canal.
- Displacement of the fluid in the lower canal is taken up by the stretching of the membrane covering the **round window**.
- Movement of the basilar membrane results in sensory stimulation. The movement distorts the sensory cells so that the sensory hairs register a mechanical disturbance on the tectorial membrane resulting in impulses being fired in the **auditory nerve**.
- These impulses travel to the brain where they are interpreted.

5.9 Structure of the nervous system

Structure of neurones

In the nervous system there are three types of nerve cells or **neurones**:

- **Sensory** – which bring impulses from the sense organs or receptors into the central nervous system.
- **Motor** – these take impulses from the central nervous system to the effector organs (muscles or glands).
- **Connector** (Intermediate or relay) – these receive impulses from sensory neurones or other intermediate neurones and relay them to motor neurones or other intermediate neurones.

❑ ***Action***
1. Draw and label a mammalian motor neurone.
   (Labels - dendrites, cell body, nucleus, axon, myelin sheath, Schwann cell, nodes of Ranvier, nerve/axon endings).
2. Describe the functions of the parts labelled.

Each cell consists of a cell body containing a nucleus and granular cytoplasm containing many ribosomes. These ribosomes are grouped together forming Nissl granules, which are concerned with the formation of neurotransmitter substances. From each nerve **cell body** arise fine branching structures, **dendrites**. These receive impulses from other nerve cells and carry the information towards the cell body. Some neurones also have a long membrane-covered cytoplasmic extension, the **axon**, which transmits impulses from the cell body. At its end, an axon divides into branches which form synapses with other neurones.
Peripheral neurones are surrounded by and supported by **Schwann cells**. In some cases, these grow around the axons of the nerve cells to form a multilayered fatty **myelin sheath**, found only in vertebrate nervous systems. This acts as an electrical insulator and speeds up the transmission of impulses. The myelin sheath has thin areas at intervals, **nodes of Ranvier**, which are important in impulse transmission.

The nerve impulse

The electrical change associated with a typical nerve impulse is very small (50 millivolts). Nevertheless nerve impulses can be recorded and measured using an apparatus which is sensitive to small electrical changes. Impulses can be picked up from the nerve through a pair of microelectrodes and fed into a **cathode ray oscilloscope**. This can measure the magnitude and speed of transmission of impulses and to analyse the pattern of impulses generated in different parts of the nervous system.

Transmission of nerve impulses

- ✓ The **resting potential** is the potential difference between the inside and the outside of a membrane when a nerve impulse is not being conducted.
- ✓ Resting potentials are typically minus values, the minus indicating the inside is negative with respect to the outside. The membrane is said to be polarised.
- ✓ The sodium-potassium exchange pumps (these are transmembrane proteins) maintain the concentration and an uneven distribution of sodium ions and potassium ions across the membrane.

- The composition of the cytoplasm of an axon is very different from that of the surrounding fluid. Potassium (K^+) and organic anions (COO^-) are higher inside the neurone while the concentration of sodium (Na^+) is higher outside. The membrane is more permeable to K^+ than any of the others.
- Since the concentration of K^+ is higher than outside they diffuse out. This outward movement of positive ions means that the inside becomes slightly negative.
- Nerve impulses are due to changes in the permeability of nerve cell membrane to K^+ and Na^+ which leads to changes in the potential difference across the membrane and the formation of action potential.
- Suitable stimulation of an axon results in change of potential across the membrane from a negative inside value of about –70mV to a positive inside value of +40mV. This change is called an **action potential**. The membrane is said to be **depolarised**.
- The action potential is the result of a sudden increase in the permeability of the membrane to Na^+. This allows a sudden influx of Na^+ which depolarises the membrane.
- A fraction of a second after this depolarisation the K^+ diffuse out and repolarises the membrane. There is an overshoot of K^+ leaving as the K^+/Na^+ pump restores the ionic balance. This is called the **refractory period** during which another action potential cannot be generated so ensuring a unidirectional impulse and limiting frequency.

❑ ***Action*** Draw and interpret an oscilloscope trace (showing an action and resting potential).

The action potential causes a small electric current across the membrane and as a portion of the membrane is depolarised, depolarisation of the next portion is initiated. There is a series of local currents propagated along the axon. The sodium pump is active all the time and behind the transmission; this pump restores the resting potential. Once the resting potential is restored, another impulse can be transmitted. As the impulse progresses the outflux of K^+ causes the neurone to be repolarised behind the impulse. The overshoot of K^+ leaving is restored by the K^+/Na^+ pump which restores the ionic balance. During the period between depolarisation and repolarisation impulses cannot pass. This is known as the refractory period.

Properties of nerves and impulses

The 'all or nothing law'

The size of the impulse is independent of the size of the stimulus.
If the strength of a stimulus is below a certain threshold intensity, no action potential is initiated. But, if the intensity of the stimulus exceeds the threshold value, an action potential *is* initiated. Any further increase in the intensity of the stimulus does not give a larger potential, i.e. the 'amount' of information sent through the nervous system is determined by the number and frequency of the impulses and not by their size.

Factors affecting the speed of conduction of the nerve impulse

Two factors are important in determining the speed of conduction of the nerve impulse.

- ✓ **Myelination** speeds up the rate of transmission of impulses by increasing the distance over which the local currents can bring about depolarisation. When axons have a myelin sheath there is no contact between the axon membrane and the tissue fluid except at the nodes of Ranvier. The result is that the impulse jumps from one node to the next, speeding the overall passage along the axon.
- ✓ The **diameter of the axon** - the greater the diameter of the axon, the greater the velocity of transmission.

Synapses

Neurones are not in direct contact with each other but are separated by tiny gaps known as synapses. The main role of the synapse is to convey action potentials between neurones. A range of drugs function by interfering with the neurotransmitters involved in synaptic transmission.

Structure of a synapse

Most junctions between neurones take the form of chemical synapses. Branches of axons lie close to dendrites of other neurones but do not touch; there is a gap of about 20um between them. When impulses are transmitted, this gap is crossed by the secretion of a neurotransmitter from the axon membrane (pre-synaptic membrane), which diffuses across the space to stimulate the dendritic membrane (post-synaptic membrane).

❑ ***Action*** Draw a diagram of a synapse.
(Label - synaptic knob, synaptic vesicle, mitochondria, pre-synaptic membrane, synaptic cleft, post-synaptic membrane.)

Synaptic transmission

The arrival of the impulses at the **synaptic knob** alters its permeability allowing **calcium ions** to enter. The influx of calcium ions causes the **synaptic vesicle** to fuse with the **pre-synaptic membrane** so releasing the **transmitter** (acetylcholine) into the **synaptic cleft**. When the transmitter diffuses across the gap (synaptic cleft) it attaches to a receptor site on the **post-synaptic membrane**, depolarising it and so initiating an impulse in the next neurone.
The postsynaptic membrane contains specific protein receptors with which the transmitter molecules combine. Once combined, protein channels open up in the membrane, allowing sodium ions to diffuse from the cleft into the postsynaptic neurone. If the membrane becomes sufficiently depolarised, an action potential is initiated in the axon of the postsynaptic neurone. **Acetylcholine**, when released, is quickly destroyed by enzymes in the synaptic cleft, so its effect is limited and the merging of impulses is prevented. If insufficient acetylcholine is released, the post-synaptic membrane will not be stimulated. The enzyme which destroys acetylcholine is called **cholinesterase**. The resulting choline and ethanoic acid diffuse back across the synaptic cleft for further use.

Another transmitter substance is **noradrenaline**. It occurs, together with acetylcholine, in the involuntary nervous system.

❑ ***Action*** Draw annotated diagrams to illustrate synaptic transmission.

Functions of synapses

- Transmit information between neurones.
- Pass impulses in one direction only.
- Act as junctions.
- Filter out low level stimuli.

Effects of drugs

Drugs which interfere with synaptic transmission can be classified into two types:

✓ Excitory drugs – which amplify the process of synaptic transmission.

✓ Inhibitory drugs – which inhibit synaptic transmission.
e.g. organophosphorous insecticides act as a cholinesterase inhibitor. Without cholinesterase, acetylcholine remains in the synaptic cleft and causes repeated firing of the postsynaptic neurone.
If the inhibitor is acting at a neuromuscular junction repeated contractions of the muscle occurs. The result is that the nervous system becomes overactive and muscles contract uncontrollably.

❑ ***Action*** Find out how psychoactive drugs, e.g. cocaine and cannabis, affect synapses in the brain.

The central nervous system

The mammalian nervous system is dual in nature. The central nervous system (CNS), consisting of the brain and spinal cord, which co-ordinate and control the activities of the animal. The peripheral nervous system, the nerves and ganglia, forms the connecting link between the organs and the CNS. Many body functions and actions are controlled by reflex actions, which involve both parts of the nervous system.

Structure and function of the brain

The brain is divided into three main regions – forebrain, midbrain and hindbrain.

Forebrain

The dorsal part of the forebrain consists of two large **cerebral hemispheres** forming the cerebrum. All the higher mental activities take place here including the interpretation of sensory information, thought processes and memory.
The **hypothalamus** is found at the base of the cerebrum. It controls the autonomic nervous system hence involuntary activities and some basic behaviour patterns. It also monitors the composition of the blood and regulates the temperature of the body.
The hypothalamus is a link between nervous and endocrine regulation.

Midbrain

This acts as an important link between the forebrain and the hindbrain.

Hindbrain

This consists of two main parts:

- The **medulla oblongata**, which contains centres controlling reflex activities such as ventilation rate, heart rate and blood pressure.
- The **cerebellum**, which is concerned with the coordination of body movements and postural responses. Impulses from the motor centres in the forebrain pass to the spinal cord via the cerebellum. It also receives sensory impulses and has the function of moderating the motor impulses to produce appropriate muscular movements.

❑ ***Action*** Draw a diagram of the brain showing the positions of the parts in bold described above.

Spinal cord

This is the portion of the central nervous system that passes through the vertebral column and from which most of the **peripheral nerves** originate. It consists of a central area of grey matter, consisting mainly of nerve cell bodies, and a surrounding area of white matter, which consists of nerve fibres. Sensory fibres of the peripheral system enter the cord via the dorsal routes. The cell bodies of the sensory fibres are found in the dorsal root ganglia, which lie alongside the spinal cord. The motor fibres leave via the ventral routes.

❑ ***Action*** Draw a labelled diagram of a section through the spinal cord.
(Labels - central canal, grey matter, white matter, dorsal root, ventral root, sensory neurone, dorsal root ganglion, connector neurone, motor neurone, effector, meninges.)

The functions of the spinal cord are to relay impulses in and out of any particular point along the cord, and to relay impulses up and down the body, including to and from the brain.

Reflex arc

- A reflex action is a rapid, automatic response resulting from nervous impulses initiated by a stimulus. They are generally protective in function.
- The following describes a typical reflex action – the nerve pathways involved in the flexion of the arm in response to touching a hot surface.

✓ Stimulus – the hot surface.
✓ Receptor – temperature and pain receptors in the skin.
✓ Sensory neurone sends impulse to spinal cord.
✓ Relay neurone connects sensory neurone to motor neurone.
✓ Motor neurone sends impulse to effector (muscle).
✓ Response – arm muscles contract and hand is removed from surface.

❑ ***Action*** Draw an annotated diagram of the nervous pathways of this reflex arc.

- The reflex arc is related to nerve pathways involved in voluntary actions via ascending and descending tracts in the spinal cord.
  With any reflex action there is a pathway for impulses to be sent to the brain via ascending nerve fibres that originate at synapses in the grey matter of the spinal cord. The brain may store this information or it may relate the information with sense data from, say, the ears. As a result of receiving this extra information impulses may be sent from the brain to modify the response. Sometimes the response is over-ridden by the brain along inhibitory nerve fibres.

Autonomic nervous system

(*This information is not on the syllabus.)*
The autonomic nervous system is divided into two parts; the **sympathetic system** which generally stimulates effectors, and the **parasympathetic system**, which generally inhibits effectors. They work **antagonistically** to each other, for example the sympathetic system increases ventilation rate, while the parasympathetic system decreases ventilation rate.

Muscles

Effectors produce the body's response to stimulation. The effectors are muscles responsible for body movement and position, and glands that secrete various chemicals.
Movements of the body are brought about by the contraction of voluntary muscles, which are attached to movable bones of the skeleton. You should look at the structure of skeletal muscle as seen with the light and electron microscopes. To understand how a muscle contracts you need to study the sliding-filament theory.

The structure of muscle

- An individual muscle consists of hundreds of **muscle fibres**. Each fibre is made up of a mass of parallel **myofibrils** running the length of the fibre. When viewed under the light microscope myofibrils show a pattern of alternate light and dark bands (known as I and A bands). Hence the alternative name for voluntary muscle is 'striped' or 'striated' muscle.

❑ ***Action*** Draw voluntary muscle as seen under the light microscope.

- As seen under the electron microscope striated muscle consists of fibres, each of which is a long, multinucleate cell. The cell wall, or **sarcolemma**, encloses several nuclei, cytoplasm and numerous mitochondria. It also contains the contractile elements, which are long protein fibres, the myofibrils. Each myofibril is divided into **sarcomeres** by cross partitions, Z lines. In each sarcomere there are two types of protein:

✓ thin **actin** filaments projecting into the sarcomere from the Z lines.
✓ thicker **myosin** filaments between them, which are fixed in position by molecular bonds.

❑ ***Action***
1. Draw the structure of a myofibril as seen under the electron microscope. [Show the arrangement of thin actin and thick myosin filaments; labels - sarcomere, Z line, A band (entire length of myosin filaments including overlap with actin), I band (actin only), H zone (myosin only), M line.]
2. Draw cross sections of the sarcomere at different points.

The neuromuscular junction

For it to contract, striated muscle must be stimulated by an impulse from a motor nerve. The point where the motor nerve meets a striated muscle is called the **neuromuscular junction** or end plate.
The structure of a neuromuscular junction is similar to a synaptic knob and the post synaptic membrane is greatly folded to form a motor end-plate.

❑ ***Action*** Draw a labelled diagram of a neuromuscular junction.

When a nerve impulse is received at the end plate the sequence of events is much the same as the mechanism of synaptic transmission.

❑ ***Action*** Describe the sequence of events which occur at the end plate.

The sliding-filament theory of muscular contraction

This theory suggests that the muscle filaments do not shorten when the muscle contracts but that they slide between one another. The theory is based on the study of electron micrographs of muscles fixed at different degrees of tension. It appears that the actin filaments, and therefore the Z lines to which they are attached, are pulled towards each other, sliding as they do so over the myosin filaments.

Contraction mechanism

- When the muscle receives a nerve impulse the sarcomeres shorten due to increased overlap of the actin and myosin strands.
- ✓ H zone shortens.
- ✓ I bands shorten.
- ✓ Z lines become closer together (i.e. sarcomere shortens).
- ✓ A band does not change in length.
- The actin filaments slide over the myosin filaments, with no change in length of either.

❑ ***Action*** Draw diagrams to illustrate sliding-filament theory.

The '**ratchet mechanism**' may be used to explain how the actin and myosin filaments slide past one another.

- ✓ There are cross bridges between the actin and myosin filaments.
- ✓ When an impulse reaches the neuromuscular junction acetylcholine is released which causes the sarcolemma of the muscle to be depolarised. This causes the release of calcium ions which activate the protein **troponin**. This displaces another protein, **tropomyosin** (which has been blocking the actin filament) away from attachment sites on the actin molecule so myosin and actin can be linked by cross bridges.
- ✓ The bulbous heads along the myosin filaments form these bridges.
- ✓ Each myosin filament has a number of bulbous heads.
- ✓ The clubbed heads of the myosin molecule swing back and forth with the breaking and reforming of bonds.
- ✓ The bulbous heads progressively moves the actin filament along, as it becomes attached and detached.
- ✓ The process receives energy by the hydrolysis of ATP. Phosphocreatine is used to regenerate ATP.

❑ ***Action*** Explain the ratchet mechanism by means of annotated diagrams.